Exploring the Galaxies

EXPLORING THE GALAXIES

Simon Mitton

CHARLES SCRIBNER'S SONS
New York

1 3 5 7 9 11 13 15 17 19 I/C 20 18 16 14 12 10 8 6 4 2

Printed in Great Britain
Library of Congress Catalog Card Number 76-42913

ISBN 0-684-14862-5

Acknowledgements

The author wishes to thank the following colleagues and institutions for supplying illustrations.

Figures: 1, 2 and 3, Michael Hoskin; 13, Adapted from a diagram by A. Sandage; 14, H. L. Johnson and W. W. Morgan; 19, Barry Madore; 21, D. S. Mathewson and V. L. Ford (Royal Astronomical Society copyright); 23, D. S. Mathewson, M. N. Cleary, and J. D. Murray (University of Chicago Press copyright); 24, Eric Becklin and Gerry Neugebauer (University of Chicago Press copyright); 25, Nicholas Scoville (University of Chicago Press copyright); 29, P. Hargrave and M. Ryle; 30, Mullard Radio Astronomy Observatory; 31, Harry van der Laan; 32, M. Ryle and M. D. Windram; 34, Mullard Radio Astronomy Observatory.

Plates: 1, Alan Stockton, University of Hawaii; 2, Dennis Downes, Max-Planck Institute, Bonn; 3, Vincent Reddish, Royal Observatory, Edinburgh; 4, 5, Royal Astronomical Society; 6, Guy Pooley, University of Cambridge; 7, Royal Astronomical Society; 8, Hale Observatories; 9, Wallace Sargent, Hale Observatories; 10, Hale Observatories; 11, Hale Observatories; 12 Wallace Sargent, Hale Observatories; 13, Hale Observatories and Mullard Radio Astronomy Observatory; 14, Hale Observatories; 15, Hale Observatories; 16, Roger Lynds, Kitt Peak National Observatory; 17, Vincent Reddish and UK Science Research Council; 18, Harry van der Laan, Westerbork; 19, Sidney van den Bergh, David Dunlap Observatory, Toronto; 20, Harry van der Laan, Westerbork; 21, Alan Stockton, University of Hawaii; 22, Halton Arp, Hale Observatories.

Contents

	Acknowledgements	*page* 7
	List of Illustrations	11
	Author's Preface	15
1	Discovery of the galaxies	17
2	Observing the extragalactic universe	34
3	Distances	44
4	Explorations of normal galaxies	59
5	Inside the galaxies	79
6	Galactic interactions	97
7	The nuclei of galaxies	109
8	Clusters of galaxies	125
9	Radio galaxies	133
10	Quasi-stellar objects	151
11	What lies between the galaxies?	167
12	The energy problem	178
13	Origin and evolution of the galaxies	187
	Index of names and subjects	201
	Index of astronomical objects	205

Illustrations

Plates (*between pages 112 and 113*)

1 The 2.25-metre telescope of the Institute for Astronomy, Hawaii
2 100-metre instrument at Effelsberg, in West Germany, the world's largest fully steerable radio telescope
3 The United Kingdom 1.2-metre Schmidt telescope stationed in Australia
4 The spiral galaxy M31 in Andromeda
5 The Trifid Nebula
6 Radio contours of M31 superimposed on the optical structure
7 M51 and its companion galaxy NGC 5195
8 The Perseus cluster of galaxies showing radio contours
9 VV172 chain of five galaxies
10 Arp 330
11 Stefan's Quintet
12 Seyfert's Sextet
13 Exploding galaxy M82 and its nuclear radio source
14 The jet of M87
15 NGC 4151, one of the brightest Seyfert galaxies
16 The nucleus of NGC 1275
17 A cluster of galaxies in Hydra photographed by the UK Schmidt telescope
18 The magnetic structure in the NGC 1265 tadpole galaxy
19 Dust clouds in the Small Magellanic Cloud, indicated by the black outlining
20 A radiograph of DA 240, the largest known object in the universe
21 The double quasar 4C 11.50 and neighbouring galaxies
22 Is quasar PHL 1226 associated with the nearby galaxy?

Line Diagrams

1 Wright's conception of the Milky Way *page* 21
2 Another of Wright's models 22
3 Model of the Galaxy made by William Herschel 23
4 Sketch of M51 made by Lord Rosse 24
5 Kapteyn's model of the Galaxy 27
6 Early twentieth-century model of the Galaxy 28
7 The Hubble sequence of galactic types 30
8 The complete electromagnetic spectrum 34
9 The synthesis of large radio telescopes 39
10 Measurement of stellar distances by trigonometrical parallaxes 44
11 The period-luminosity relation for Cepheids in the local group of galaxies 49
12 Comparison of methods of measuring distances in the universe 56
13 Sandage's colour-magnitude diagram for cluster stars 60
14 The Hertzsprung-Russell diagram for nearby stars 62
15 Rotation curves of spiral galaxies 71
16 Correlation of galactic colour with morphological classification 75
17 Colour-colour diagram for galaxies 76
18 Spiral structure of the Galaxy deduced from radio astronomical observations 82
19 Location of objects of known distance in the Milky Way 85
20 Polarization of electromagnetic radiation 86
21 The magnetic field of the Galaxy 87
22 Location of globular clusters 95
23 The Magellanic stream 104
24 Map of the galactic centre at 10 μm 111
25 Model of the galactic nucleus 114
26 The local group of galaxies 127
27 Typical radio galaxy spectra 137
28 Synchrotron radiation 139
29 High-resolution map of Cygnus A radio galaxy 141
30 Radio map of 3C 465 radio source 144
31 The largest object in the universe, 3C 236 145
32 Galaxies and radio sources in the Perseus cluster 147
33 The variation of redshift with velocity 154

34 Radio map of a quasar, 3C 47 *page* 156
35 The microwave background 188
36 Formation of galaxies from instabilities 191

Tables

1 The electromagnetic spectrum 35
2 Distance determination 57
3 The classification of stars by spectral type 63
4 The UBV filter system 75
5 Membership of the two basic stellar populations 80
6 Molecules found in interstellar clouds 113

Author's Preface

The exploration of the universe must be regarded as one of the great intellectual achievements of our civilization. Astronomers can now take us back to an early epoch in the evolution of the universe by means of large and sensitive telescopes. Our knowledge of the scale of the cosmos and of its origins derives largely from observations of objects beyond the Milky Way. These celestial bodies include quasars, radio galaxies, and active galaxies as well as the normal spirals and ellipticals. Our understanding of the extragalactic universe has increased greatly now that radio, infrared, and X-ray measurements can be made fairly readily. The opening of these invisible parts of the spectrum has led to an explosive growth of research activity on extragalactic objects.

In this book I have attempted to explain the basic properties of galaxies and the main trends of present research in non-technical language. The second half of the book is concerned with an exposition of the major problems of extragalactic research that confront the professional astronomer.

Many colleagues assisted me while I was writing this book. I particularly thank Donald Lynden-Bell, Jacqueline Mitton, Patrick Moore, Martin Rees, John Shakeshaft and Peter Stubbs for their interest and encouragement, as well as the many people who provided illustrations. A monumental amount of typing was expertly dealt with by Gertrude Pardoe.

Simon Mitton
Institute of Astronomy, Cambridge, England
May 1975

1 · Discovery of the galaxies

1.1 Exploration of the universe

Man's exploration of his environment has now penetrated to the furthest depths of the universe, almost as far back as the beginning of time itself. With the aid of powerful telescopes the professional astronomer can look out practically to the limits of the cosmos. From these furthest boundaries radiation takes many billions of years to journey to our planet, with the result that the astronomer is looking back across aeons of history when he surveys the distant universe. By careful examination of the matter in the universe, particularly the material at very great distances, it should be possible to deduce the past history of the universe and of the objects within it.

Throughout our vast universe there are countless isolated star systems, separated by immense distances. Two centuries ago, before their true nature became evident, these starry concentrations were termed nebulae, on account of their misty appearance when viewed by eye through a simple telescope; *nebula* is the Latin for cloud. Today we refer to the wheeling star systems that make up the 'atoms' of our universe as galaxies. Investigation of the galaxies has, in the past half century, resulted in the most dramatic era that astronomy has ever witnessed; at the present time the advances seem to come at an ever increasing rate, for contemporary scientists can probe more than ever before, with their radio, infrared, optical and X-ray telescopes. The cosmologist Dennis Sciama (Oxford) has described the exploration of the universe as 'the greatest intellectual adventure of the mid-twentieth century'. Discoveries since 1960 alone have resulted in a revision of our conception of the universe that is without precedent in the history of science. Crucial to these events have been careful studies of galaxies and the remote regions between the galaxies, for they hold a key to partial exploration of the physical universe.

1.2 Early beginnings

Man's earliest questionings about the sky are buried deep in prehistory; we may be fairly sure that long ago the sun and moon received far more respect and attention than did the stars. In neolithic times (*circa* 3500–1500 B.C.), for example, sophisticated observatories for solar and lunar work were erected throughout Britain and parts of northern France. There is also a suggestion that these primitive people followed the rising and setting of the bright stars. Today half-ruined stone circles are all that remain of a scientific effort that spanned a millennium or more. One can only admire the hardy folk who erected the megalithic circles, such as Stonehenge, to help with their construction of a calendar and the prediction of eclipses.

From the standpoint of galactic studies, the first person of note in the classical world was Democritus, of the sixth century B.C., who speculated that the universe was populated with an unending series of worlds like the Earth. Two thousand five hundred years elapsed before a shred of evidence could be advanced to support this philosophy. For twenty-one centuries after Democritus the study of the grand design of the starry universe (as opposed to that of the planetary system only) remained dead.

With the Renaissance, science emerged from the Dark Ages and men questioned the classical authors. The first astronomer of consequence in Western Europe was Nicolas Copernicus (1473–1543), who placed the sun at the centre of our planetary system. The Earth he set in daily rotation to account for the apparent motion of the sun and stars. His model opened the possibility that the stars could be distant suns, as opposed to points of light on a solid crystalline sphere that rotated about the Earth. Giordano Bruno took the next giant leap forward when he propounded, in the sixteenth century, that the universe had no limit or edge. Furthermore, he postulated the existence of other suns, planets and lifeforms. His work implied that the stars shine only dimly in the sky on account of their great distance, compared to the sun, from the Earth.

The whole of astronomy changed dramatically once Galileo Galilei got down to the serious business of inspecting the heavenly bodies. He found that a suitable combination of a concave and a convex lens rendered distant objects larger than they appeared to the naked eye. Certainly he was the first man to train an optical telescope on the heavens and describe in writing what he saw. He discovered the mountains and craters on the moon, observed sunspots, and, of great importance, the system of four large satellites around the planet Jupiter. Here was a solar system in miniature, living proof that the planets must in their turn circle the sun.

Galileo looked towards the fragmented clouds of the Milky Way—that distant band of light that girdles the heavenly sphere. With great excitement he wrote down his findings in his book *The Sidereal Messenger*:

> I have observed the nature and material of the Milky Way. With the aid of the telescope this has been scrutinized so directly and with such ocular certainty that all the disputes which have vexed philosophers through so many ages have been resolved, and we are at last freed from wordy debates about it. The galaxy is, in fact, nothing but a congeries of innumerable stars grouped together in clusters. Upon whatever part of the Milky Way the telescope is directed, a vast crowd of stars is immediately presented to view. Many of them are rather large and quite bright, and the number of small ones is quite beyond calculation.

Galileo's telescope opened up a new era in astronomy: he demonstrated that the sun lay at the centre of the solar system; he showed that the Milky Way formed a gigantic starry cloud deep in space. Galileo started the great quest to understand the universe at large.

1.3 The nebulae

Scattered through the sky there are several hazy patches which are just visible to the naked eye on a clear night. A small telescope or binoculars will reveal several dozen clouds of light. In the Orion constellation a soft glow is just visible below the hunter's sword. The Andromeda constellation also contains a well-known patch of light which looks larger than the stars. These cloudy patches are the nebulae mentioned earlier. Galileo inspected some of the nebulae with his telescope and he resolved a proportion of them into a tight mass of stars. Generally speaking, however, no one was very interested in the nebulous patches, and by 1700 only a dozen or so had received a cursory glance.

The turning point came in 1705 when the Englishman Edmond Halley published his book on comets. He claimed that a series of bright comets which were sighted in 1531, 1607 and 1682 could be explained by a single object which returned about every seventy-six years. He demonstrated that comets were part of the sun's family, moved in accordance with Kepler's Laws of planetary motion. About this time great hunts for comets commenced, for the recognition of a new comet brought instant fame to its discoverer. Nightly the comet seekers would sweep the sky in a quest for new faint patches of light, characteristic of a comet, but not unlike the fixed nebulae. When a new one was found the observer would mark it on a sky map and hope that it would show some motion by the next evening.

The Frenchmen Charles Messier and Pierre Méchain spent thirty years hunting new comets. In the course of their endeavours they found that the fixed nebulosities caused much confusion. To keep track of these interlopers they drew up a catalogue of 103 nebulae and listed their positions. Only the brightest nebulae appear in Messier's catalogue, and it therefore became a natural starting point that eventually led astronomers to study the prominent nebulae of the Milky Way and beyond. Objects in the Messier catalogue are referenced by the prefix M followed by the number of their catalogue entry. Thus M31, the great nebula in Andromeda, is the thirty-first source listed by Messier. Ironically the nebulae were nothing more than an irrelevant side issue for the comet hunters, and yet they carried clues to the gross composition of the universe.

About this time the Englishman Thomas Wright was actively engaged on astronomy, and in 1750 he published an important treatise *An Original Theory*, or *New Hypothesis of the Universe*, which was more or less ignored by his contemporaries. This was a pity as it contained some very fruitful ideas on the structure of the universe.

Thomas Wright of Durham developed the idea of Democritus and Galileo, that the Milky Way is composed of countless myriads of distant suns. He gave the following explanation of the appearance of the Milky Way and account of the misty patches of light. First, he rightly supposed that the Milky Way is but one of many similar objects throughout the universe, in which the stars might form themselves into coherent groups that are sprinkled throughout space. He pointed to the misty nebulae as visible evidence that systematic arrangements of very distant stars did exist in the universe. Wright's second point is also essentially correct: he speculated that the Milky Way appears as a bright band in the sky because it is a flat disc of stars which we view from the inside. His book gives a diagram of our Milky Way with the stars arranged in a slab (Figure 1). He was the first to suggest the disc model of our Galaxy, the model that is now accepted by all astronomers.

Within a few years the great philosopher Immanuel Kant further extended Wright's ideas. While at the University of Konigsberg, Kant read a garbled account of Thomas Wright's book in a local newspaper. During the next four years Kant beavered away on unlocking the mysteries of the Milky Way and cloudy star patches. Kant accepted the disc model for the distribution of stars in our

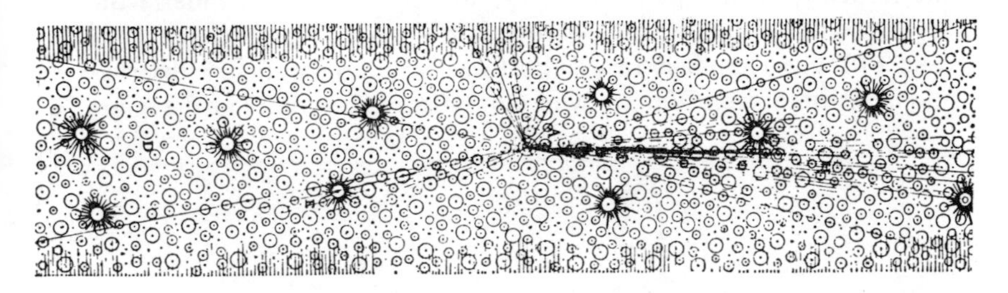

Fig. 1. Wright's representation of the Milky Way by a flat disc of stars

Galaxy, and proposed that the shape of the Milky Way, and indeed the distant nebulae, might be determined by rotation. Looking at the oval nebulosity in Andromeda, he argued that it must be a great mass of fixed stars, the whole rotating about the centre of the ellipse. The outstanding contribution made by this philosopher was his proposal that many of the nebulae should be identified as remote objects similar to the Milky Way. This concept remained essentially unaltered for the next century and a half; the idea became known as the theory of island universes—a theory which postulated that starry nebulae were scattered through space, to infinite distances.

1.4 The Herschel family take up stargazing

William Herschel probably spent more hours looking through telescopes than any man before or since. Paradoxically, he did not show the slightest interest in the night sky until the age of thirty-five.

Nevertheless he stands alongside Galileo Galilei and Isaac Newton as an outstanding contributor to the modern universe. William Herschel, a Hanoverian, was born into a family of musicians in 1738, and he soon joined a regimental band. Subsequently he decided to go to England and establish himself in London in November 1757. Later his sister Caroline and brother Alexander joined him, and he earned his keep from musical activities.

William Herschel purchased a copy of a leading astronomy textbook in 1773. He was so thrilled to read of the discoveries that had been made by the telescope that he determined to build one himself. He studied a book on optics, and in no time at all had set up a workshop in the family home at Bath to build immense telescopes for surveying the sky. Herschel's enthusiasm grew at a great rate as he turned his telescope on the sky, and he spent much time in the construction of larger instruments. Thus it was that between 1773 and 1782 the Herschels made the transformation to being observational astronomers. William was appointed Astronomer at the court

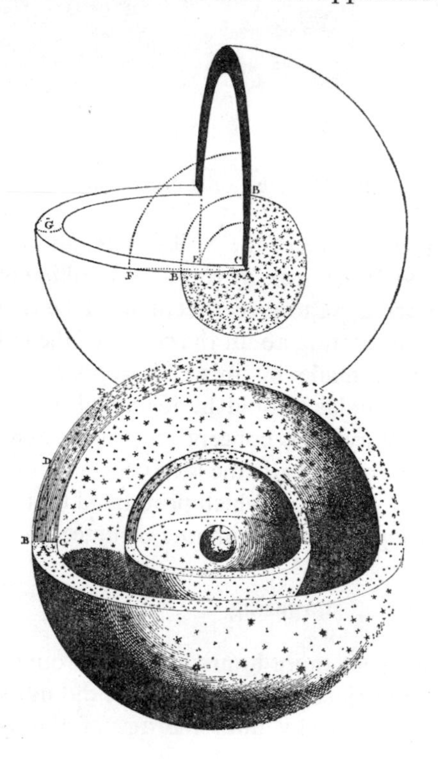

Fig. 2.
Another model of the distribution of stars, due to Wright

of George III and Caroline became known in Europe as an astronomer in her own right. Herschel's diligence was rewarded by his discovery of a new planet, Uranus, on 13 March 1781.

To the modern astronomer, however, Herschel's lasting contribution was his catalogue of nebulae. His telescopes were superior to all others; his largest had an aperture of 48 inches and a focal length of 40 feet. Inspired by the Messier catalogue, he decided to study the nebulae and thence find the grand design for these objects. He determined their positions, dictating the angle of his telescope and a brief description of the appearance of each nebula to Caroline. Within seven years they had found 2000 previously uncatalogued

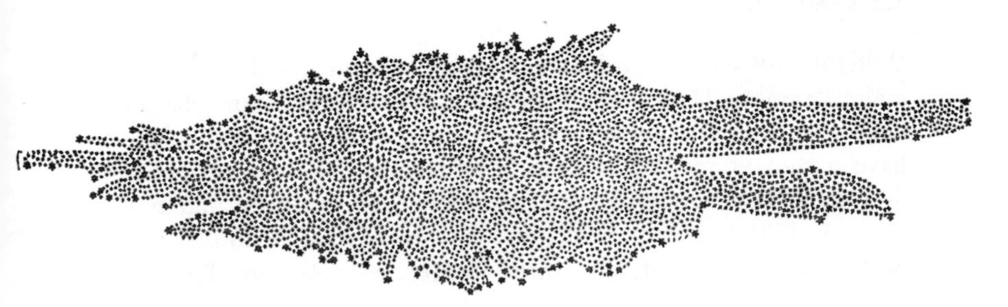

Fig. 3. William Herschel's model of the Milky Way, deduced from counts of the star density

objects. Many of his drawings and models (Figure 3) were published by the Royal Astronomical Society of London, and he enthusiastically wrote

> I have seen double and treble nebulae, variously arranged; large ones with small seeming attendants; narrow but much extended, lucid nebulae or bright dashes, some of the shapes of a fan, resembling an electric brush, issuing from a lucid point, others of cometic shape, with a seeming nucleus.

William's son, John Herschel, and J. Dreyer continued the work of cataloguing the nebulae during the nineteenth century. After completing the work on the northern sky, John Herschel moved to the Cape of Good Hope in 1833, and set to work on the southern nebulae.

All this work of cataloguing had shown, from the very start, that the nebulae do not form a homogeneous group of objects. William

Herschel tried to describe the motley assortment he found in the nebular rag-bag. Some were tight clusters of stars, others were great gas clouds; then there were the ones that looked like small glowing discs—the planetary nebulae. Ultimately two main types of nebulae became recognized: galactic and extragalactic. In the former class are the great star clusters of the Milky Way, along with the planetary nebulae and gas clouds of the Galaxy. The other category refers to those nebulae that are not part of the Milky Way; small, symmetrical objects scattered uniformly over the sky and called extragalactic nebulae. These are the island universes of Immanual Kant.

1.5 Cosmic whirlpools

William Parsons, the third Earl of Rosse, was intrigued by machinery and astronomy. He set about the construction of gigantic telescopes. One with a 36-inch mirror performed successfully, so he decided to have a shot at making a 72-inch telescope with a focal length of 54 feet. The 6-foot mirror blank took a year to grind and polish. By early 1845 the whole instrument was ready for action; it was to be the world's largest, and most dangerous, telescope for several decades.

Lord Rosse started to work through Herschel's list of nebulae, and found that many of them could be resolved into stars. He turned his telescope on M51; Messier had described this as a double

Fig. 4. Lord Rosse's famous drawing of the spiral structure in M51. Compare this sketch with the photograph in plate 7

nebula with two 'atmospheres' touching. Lord Rosse realized, on examining M51 several times, that it displayed a spiral pattern. He soon found other examples of this whirlpool structure among the nebulae, and he drew many of them with great care. His sketch of M51 clearly shows the spiral pattern and the additional bright nebula to one side of the main structure (Figure 4). Rosse firmly believed that these were vortices of stars, and the conclusion that they were rotating therefore seemed inevitable.

The discovery of spiral structure in the external nebulae was very significant. It excited the astronomers of Rosse's time because Pierre Laplace, the great French mathematician, had theorized that the sun and solar system had condensed from a whirlpool of cosmic gas. The spiral nebulae confirmed that such structures did exist in the cosmos; even today the precise mechanism that determines the spiral structure is not fully explained.

1.6 Power to the astronomers' eyes

In the latter part of the nineteenth century, the development of photography and spectroscopy set astronomy on a new course. John Herschel had in fact dabbled in photography, but the technique never reached a sufficiently advanced state for application to celestial work during his lifetime. Towards the turn of the century plates with dry emulsion became available. The distinguished American astronomer Henry Draper photographed the brighter nebulae; his methods caught on, and soon astronomers throughout the world were able to obtain permanent and precise images of celestial objects. The nebulae could now be studied in far greater detail as the photographic plate could record much fainter structures during long exposures than the human eye is capable of.

The spectroscope is an instrument which disperses a ray of light into its component colours. Early spectroscopes used prisms to reveal the rainbow of colours in sunlight. Joseph von Fraunhofer, a German optician, examined the light from the sun and found that its spectrum contained hundreds of dark lines. On looking at the stars, astronomers found that refracted starlight also had dark lines in its spectrum.

In the middle of the nineteenth century physical chemists showed that each of the chemical elements has its own characteristic spectrum. This can be regarded qualitatively as a fingerprint, impressed in the electromagnetic radiation, from which the nature of the element can be deduced. If an element is excited, perhaps by heating or passing an electrical discharge through it, energy is radiated as a series of emission lines. These are bright enhancements of the light at particular frequencies, which vary from element to element. In stellar astrophysics one is more commonly dealing with absorption lines. These are impressed into the light from a star when the radiation has to pass through a layer of gas that is cooler than the region where the optical radiation is emitted. In the case of stars the absorption lines are present in the spectrum because the atmosphere is cooler just above the bright surface layers. By studying the absorption-line spectrum for a star in sufficient detail it is possible to determine the relative abundances of the elements composing its atmosphere. Gustav Kirchhoff and Robert Bunsen identified more than twenty elements in the atmosphere of the sun by analysing the solar spectrum and matching the lines to elements previously investigated in the laboratory. They pioneered the technique that is now used for finding the chemical composition of objects deep in the universe.

A British pioneer in the analysis of starlight, Sir William Huggins, attacked the nebulae with the spectroscope and got a surprise. Some of the nebulae had a spectrum that contained bright lines, rather than dark ones. From an examination of five dozen nebulae and clusters of stars, William Huggins found that one-third had emission-line spectra. Most of these are planetary nebulae, a class of very hot gaseous objects, within the Milky Way. Huggins also recorded that the great nebula in Andromeda had a spectrum typical of myriads of superimposed stars. Thus the spectroscope could show which nebulae were gaseous (bright emission-line spectrum) and which were stellar (continuous spectrum crossed by absorption lines); consequently it soon became an important tool in the quest for the distant galaxies.

1.7 Order from chaos

At the dawn of the present century, observational astronomy was in an appalling state of chaos for several reasons. During the nineteenth century professional observatories had become established in many lands. Each observatory had projects to compile their own star catalogues. Since no international body existed to co-ordinate the research of different institutions, it was possible for the same star to be listed differently in several publications. The archives of many of the older observatories still contain large collections of these unco-ordinated nineteenth-century star catalogues. It was not possible to deduce the architecture of the local stellar system and Milky Way while such a hit-and-miss approach existed for positional astronomy.

On another front the astrophysicists were at loggerheads with the geologists. By use of classical thermodynamics, Lord Kelvin had shown that the age of our sun could only be 100 million years, so that the Earth was presumably younger still. This estimate assumed that the sun derived its heat energy from the gradual contraction that is caused by the force of gravity. The Earth scientists, however, deduced that our planet is billions of years old. Their arguments were based on studies of the sequences within rock strata and the fossil life embedded within rocks.

A third impasse concerned the general problem of distance in astronomy. How far away were the stars and nebulae? At the time the distances of some 100 nearby stars could be found by conventional surveying and triangulation techniques, as explained further in Chapter 3. But how could man then measure across the void to more distant objects? His interstellar surveyor's chain would stretch only 100 light years or so. Related to this was the question of the nature of the nebulae, especially the location of the various spindly, spiral and elliptical patches of light found all over the sky, except in the Milky Way where gaseous nebulae prevailed.

The Dutch astronomer Jacobus Kapteyn (1851–1922) imposed order into the star work. A pioneer of the statistical approach to astronomy, he attempted to fathom the sun's starry neighbourhood by making counts of the density of stars as a function of their distance

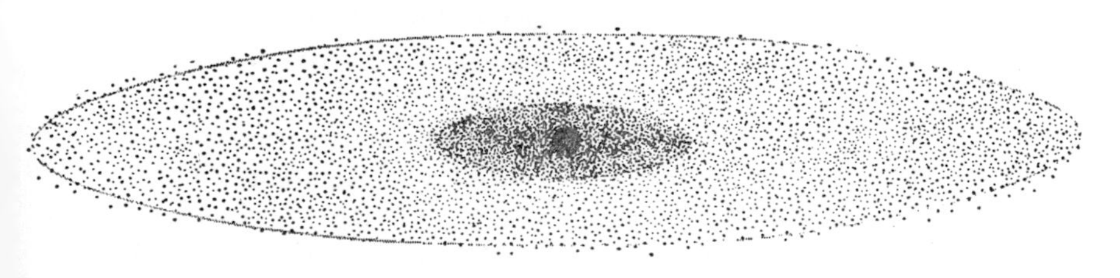

Fig. 5. Kapteyn's star counts placed the sun (solid blob) at the centre of the Galaxy, which was thought to be 1,200 light years in diameter

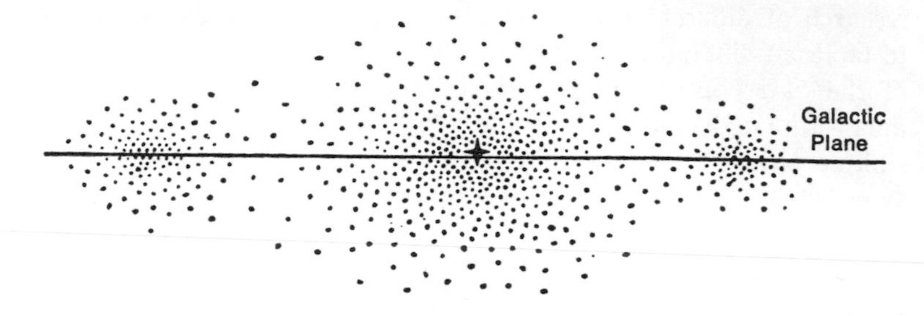

Fig. 6. During the 1920s it was still considered that the sun (large star) lay at the centre of the Galaxy. This rendition is by Sir Arthur Eddington

from the sun. From this work Kapteyn deduced the distribution in space that the stars must have in order to reproduce the observed distribution. Kapteyn's model of our Galaxy was a small disc (1200 light years diameter) with the sun at its centre (Figures 5 and 6). At its time, however, this model was only one of several.

The great breakthrough came from astronomers at the Harvard College Observatory. A research assistant at Harvard, Miss Henrietta Leavitt, found the key that finally unlocked the enigma of nebular distances. She concentrated on the properties of variable stars, especially the relationship between brightness and variation period. By studying variable stars in the Small Magellanic Cloud she found, in 1912, an important relationship for a group of variables known as Cepheids: the brightness of Cepheids in the Magellanic Clouds increased with the time taken for one complete cycle of variation.

A European astronomer, Ejnar Hertzsprung, developed Leavitt's discoveries. He determined a rough estimate of the average intrinsic brightness of Cepheid variables in the neighbourhood of the sun. By comparing this true brightness with the apparent brightness of variables in the Magellanic Clouds he concluded that the latter were much fainter because they are a great distance from the sun. He estimated this as 3000 light years. This research showed that Cepheids could be used as distance indicators: the observed period betrayed their absolute or true brightness and the ratio of apparent brightness and absolute brightness was a measure of distance.

Harlow Shapley of Harvard seized on this method of using variable stars to deduce distances and applied it to clusters of stars in the Milky Way. When he had determined the distances to clusters he worked out the three-dimensional distribution. The result was a model for our Galaxy at variance with Kapteyn's: the sun lay right to one edge of a disc 8000 light years in diameter. Who was right—Kapteyn or Shapley? Fierce controversy ensued, and Shapley's model, now known to be essentially correct, did not gain general acceptance until the 1920s.

The next problem was the question of spiral nebulae. Harlow Shapley regarded these as gaseous objects not particularly distant from the Milky Way, a conviction not shared by all observers. The answer came from the great 100-inch reflector on Mount Wilson, then the largest telescope in the world. Edwin Hubble succeeded in distinguishing individual stars in the great Andromeda spiral M31. The nearest spiral nebula consisted of swarms of stars! The extragalactic hypothesis—that these nebulae were island universes like the Milky Way—immediately went from strength to strength. Next Hubble identified some pulsating variables in M31. Now he was really onto a great discovery. The apparent brightnesses of the stars were measured, and the periods determined. The periods gave the absolute brightness, and thus the distance could be found. For M31 it exceeded a million light years! Detailed work on exploding stars in M31 confirmed that it was at a very great distance. The extragalactic universe had arrived. The spiral nebulae were great galaxies like our Milky Way, a major component of the universe.

1.8 Extragalactic taxonomy

Once he had finally established the extragalactic nature of the nebulae, Edwin Hubble undertook a systematic study of their structural features. For a large class of objects, such as the galaxies, one way to do this is to arrange the objects in various sub-classes and groups, according to their most outstanding features. Once this is done the most conspicuous members of each group can be singled out for detailed study. So, this task in astronomy is rather like elementary botany: the astronomer is trying to draw up family trees relating to different types of object in the hope of discovering how the various species might be interrelated. Hubble's classification (Figure 7) is the simplest which has been devised for galactic work, and it delineates the major patterns among the several hundred objects on which he worked. There are three broad categories: spiral galaxies, elliptical galaxies and irregular galaxies.

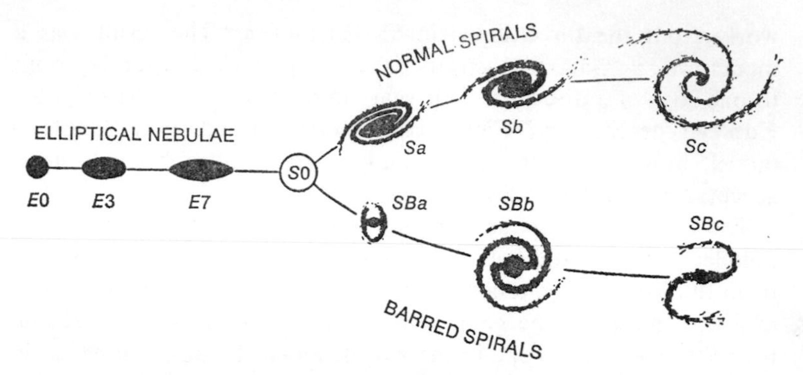

Fig. 7. Hubble's simple classification scheme has stood the test of time. It is not an evolutionary sequence

Spiral galaxies are typified by the Andromeda Nebula M31, a face-on spiral M33, and M81. Our own Milky Way system looks like a spiral from the outside. Astronomers were long intrigued by the spiral arms: do they trail behind or lead off in front? It is now known that the arms are indeed trailing, and so falling behind as the rest of the galaxy rotates about its central core. The central areas of spirals show a variety of forms. In M31 the nucleus is a bright predominant bulge, whereas M33 possesses a relatively inconspicuous condensation at its centre. The spirals break down into two subgroups: normal and barred, and there is a sprinkling of objects that are intermediate between the two types. Generally the *normal spirals* are galaxies in which two arms emerge smoothly from opposite sides of the central region. Thence they wind out along paths and merge with intergalactic space. *Barred spirals* are so named because they have a bright bar, across the central area, that links the inner ends of the spiral arms; the latter appear to extend out from the bar. One fascinating feature of these luminous bridges is that they rotate as if they were a single solid object, whereas they are in fact composed of individual stars.

Elliptical galaxies are very common, comprising perhaps 80 per cent of the galaxies visible through a large telescope. Their name derives from the smooth image, in the shape of an ellipse, that they leave on a photographic plate. A whole range of shapes is found, from spherical galaxies, through ellipsoidal figures, and on to flattish lens-shaped objects. Ellipticals are highly concentrated, and recent research shows that they smoothly increase in their brightness from the periphery up to the centre. Two small elliptical galaxies are satellite companions to M31; one of these is spheroidal and the other an ellipsoid. Elliptical galaxies come in a great range of sizes, from dwarf (10^6 times the sun's mass) up to supergiant (10^{12} times the sun's mass), and the larger ones are occasionally powerful emitters of radio waves.

Finally, in Hubble's classification, the *irregular galaxies* comprise a few per cent of the total. Their name is really a catch-all for objects that do not fit the spiral or elliptical nomenclature. These are distorted or irregular in outline. They are now the subject of intensive research, because the rag-bag of irregular objects is actually a rich hunting ground for exploding galaxies, peculiar objects, tidally-interacting galaxies and young dwarf galaxies. The label irregular is associated with a multitude of short-lived, and perhaps abnormal, evolutionary stages. A handful of astronomers have scoured the heavens for these celestial oddities, and they have published photographic atlases that illustrate a variety of tortuous forms. The names of the American astronomers Halton Arp and Fritz Zwicky and the Russian B. Vorontsov-Velyaminov are particularly prominent in this field of extragalactic research.

When giving a brief description of a particular galaxy, it is useful to have an abbreviated code for indicating the overall structure that the galaxy possesses. Edwin Hubble evolved a shorthand scheme that is still useful for some branches of research, although it has largely been superseded by more ambitious schemes. Nevertheless it has the merit of great simplicity.

Hubble distinguished the elliptical nebulae purely from their geometrical shapes. Spherical galaxies were called E0 and cigar-shaped ones (about three times as long as they are wide) were designated E7; E1 to E6 were then used for galaxies falling between the extremes, higher numbers denoting greater eccentricities. It is important to realize, as did Hubble, that a scheme of this type is little more than a convenient filing system. This is because a galaxy shaped like a lemon will appear as type E5 from one side, and may look like E0 when viewed end-on. Intermediate projections give shapes between E0 and E5. Because the orientation in space changes the code number, we do not expect this particular classification scheme to turn up profound properties of galactic structure, form and evolution. Hubble himself was interested in whether any truly spherical galaxies existed, and he was able to show statistically that there are too many E0 for them all to be projections of ellipsoidal galaxies, so that some are presumably spherical.

Classification of the spiral galaxies is more tricky as they do not have a convenient geometrical shape. The designations S and SB distinguish normal and barred types respectively; to these Hubble added a modifying letter a, b, or c. Types Sa and SBa have a large, bright, nuclear region and spiral arms that are tightly wound. Proceeding to Sb and SBb, the nucleus is less pronounced, and arms open out more, often possessing a few knotty condensations of stars and gas. Finally, the Sc and SBc galaxies have an inconspicuous nucleus, and large, wide arms that contain many condensations. Our own Galaxy is probably Sb and the Triangulum galaxy M33 is Sc.

To what extent does a sequence such as this inform us on the evolution of galaxies? In his original work Edwin Hubble definitely cautioned against accepting his galactic family tree as a true progression from youthful to ancient galaxies, although he did not reject this as being a possibility. As a matter of fact, we know for certain that galaxies do not start as ellipticals, then turn to spirals, and finally fall to pieces as irregulars, as Hubble's tuning-fork diagram might imply. Such diagrams are helpful for the visualization of the major morphological patterns present in galactic structures, but they must not be abused.

Classification schemes based on shape can obviously be devised to almost any degree of complexity, and Hubble's original proposals have been refined and extended several times. A three-dimensional scheme introduced by Allan Sandage and Gerard de Vaucouleurs is currently in widespread use. This includes letters in the coding to indicate more precisely where the inner arms are joined to the central regions of the galaxy. Many characteristics such as the colour, extent of the nucleus and the distribution of stars, gas and dust vary continuously through the sequence. Yet another way of organizing the information is to consider the spectra of different galaxy types. At the Yerkes Observatory W. W. Morgan has shown that the spectrum is correlated with the general shape of a galaxy.

Presumably schemes that are based on more easily measurable quantities such as colour or brightness distribution will be introduced in the future, and thus replace the present subjective schemes. For the time being these will continue to serve usefully as a convenient shorthand for giving brief descriptions of individual galaxies.

Recent developments show how galactic taxonomy can be overtaken by new discoveries. A category that Hubble did not mention, and only now thought to be of considerable interest, is that of compact galaxies. These are outwardly similar to ellipticals, except that they have a rather smaller size, one second of arc or less, a high density and a starlike nuclear region. During the 1960s Fritz Zwicky catalogued faint galaxies at the Hale Observatories. He circulated lists of the compact galaxies among his colleagues, and subsequent research has shown that the compact objects have a variety of properties. Some of them appear to be very young indeed, being composed of objects only a few million years old.

Another extragalactic population which has excited great interest is that of quasars. They are characterized by their very small size (they look like single stars, not galaxies, on photographs) and high brightness; they are often bluish in colour, and appear to be at the remote limits of the universe. It is probable that no class of objects in extragalactic space has attracted so much heated argument in the past decade. Many of their properties have defied an explanation, and mention of them brings us to the frontiers of modern research.

2 · Observing the extragalactic universe

2.1 Electromagnetic radiation

In comparison with the majority of scientists, astronomers labour under a fundamental disadvantage because they have no direct influence over experimental conditions, and cannot repeat experiments under a range of conditions. Almost all the information available to astronomers comes as radiant energy from celestial objects. This radiation is analysed in a variety of ways in order to infer the properties of the universe and the matter contained within it.

Electromagnetic radiation, as the radiant form of energy is called, exists in a variety of guises because it spans an enormous range of frequencies: radio, infrared, visible and ultraviolet waves, as well as X-ray and gamma-rays are all examples of electromagnetic radiation. These waves all travel in a vacuum at the same velocity: 299,793 km/sec; this quantity is often called the speed of light.

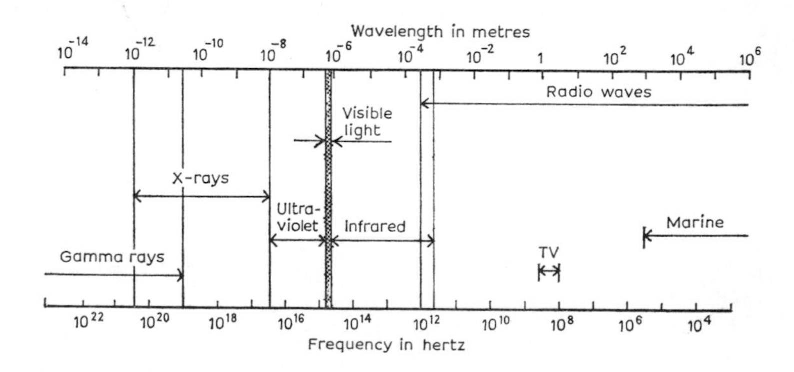

Fig. 8. Major features of the electromagnetic spectrum

TABLE 1

The electromagnetic spectrum			
Wavelength	*Photon energy or frequency*	*Type of radiation*	*Typical detectors*
			Geiger counters
10^{-5} Å	1,240 MeV		Scintillators
10^{-4} Å		Gamma rays	Nuclear emulsions
10^{-3} Å	12.4 MeV		Proportional counters
			Spark chambers
10^{-2} Å			
10^{-1} Å		X-rays	
1 Å = 10^{-8} cm	12.4 keV		
10 Å			Photography, photoelectric detectors
100 Å	124 eV	Ultraviolet	Telescopes, spectrographs
1000 Å			and spectrometers
10,000 Å = 1μ	1.24 eV	Visible	Photoconductive detectors,
10μ			radiometers
100μ	0.012 eV	Infrared	
1000μ = 1mm			
10 mm = 1 cm	30,000 MHz	Radar	
		Radio	
10 cm		UHF	
			Radio telescopes
100 cm = 1 m	300 MHz	FM	
10 m		Shortwave	
100 m	3 MHz		
1000 m = 1 km	300 kHz		
10 km			
100 km	3 kHz		
1000 km			

eV = electron Volt — Å = Ångström
Hz = Hertz = cycles per second — μ = micrometre = 10^{-6} m

For practical purposes the distinguishing character of the different radiations is wavelength. Radio waves have wavelengths from millimetres (microwaves) right up to kilometres (long waves). For infrared radiation the micron (10^{-6} m) is usually used as the unit of wavelength. For light two units are in common use; they are the nanometre (nm = 10^{-9} m) and the Ångström (Å = 10^{-10} m). In astronomy it is still more common to use Ångströms, although nanometres are in general use in much of physics. X-rays cover a wavelength range from about 10^{-8} to 10^{-11} m, and gamma-rays have wavelengths shorter than 10^{-11} m (Table 1, Figure 8).

Because electromagnetic waves have such a broad span of wavelength, a great variety of techniques has been developed to detect the radiation. The human eye is sensitive to only a very narrow range, from roughly 4000 to 7000 Å; our bodies can detect radiation just below 7000 Å as radiant heat. The visible and near-visible radiations can also be recorded on photographic plates. Radio waves from space are collected by radio telescopes of various types. Electrons in the aerial, or antenna, are excited by the extremely weak flux of radio waves focused on it, and this signal is then amplified by very sensitive receivers. For X-rays geiger counters or solid-state detectors are used; in these devices the arrival of a high-energy photon triggers an electronic response in the detector, and this response is then amplified and recorded.

2.2 Optical telescopes

The first telescopes to be invented, probably in the Low Countries in the early seventeenth century, were refractors. They use lenses to gather the light from distant objects and form a magnified image. In 1671 Isaac Newton made the first reflecting telescope, using a concave lens to focus the light. He developed this alternative instrument, the basis of all large telescopes of modern times, in order to obviate some of the inherent problems of lenses, especially the chromatic aberrations.

Larger telescopes have two principal advantages over smaller ones. In the first place the larger the telescope the more light it will collect for concentrating onto an instrument or a photographic plate. The world's largest refracting telescope has an aperture of just over 1 metre; a lens any larger than this would bend intolerably under its own weight and also absorb a considerable amount of light on account of its thickness. Consequently all the giant instruments used in extragalactic research are reflectors (Plate 1). There are now several with apertures of 4 metres: in Australia, Chile and at Kitt Peak. The world's largest reflector is the 6-metre reflector in the Crimea, and the biggest in the U.S.A. is the 5-metre instrument of the Hale Observatories. Larger telescopes allow us to see further into space because they can capture sufficient of the radiation from very faint objects to excite photographic emulsions or electronic detectors. In addition to the light-gathering power a second advantage of bigger telescopes is that they have a higher resolving power. Many astronomical investigations call for photographs showing very fine detail, and larger instruments produce sharper images because the diffraction of light within the instrument becomes less serious. Unfortunately, this effect is largely cancelled out by the turbulent motions in our own atmosphere. These cause random refractions in the light from stars, which is why they seem to twinkle, and lead to blurred images. It is rare or impossible even for the large telescopes to attain their theoretical performance for this reason. It is essential to have large telescopes operating from platforms in deep space in order to overcome the atmospheric disturbance.

Other trends are making it increasingly difficult for large telescopes to conduct essential extragalactic research. One of the most worrying is light pollution. In the cities close to many of the world's famous observatories the use of electric lighting in streets is increasing at a rate of 10 per cent per year. This has led to a serious degradation in observing conditions because photographic plates become saturated by the light from the cities before they have recorded faint astronomical images. The major observatories in continental North America will be useless for studies of faint objects by 1985–90 unless these alarming trends can be reversed. Although the new observatories in the Andes and on the Hawaiian islands are not affected to the same extent, the only satisfactory long-term solution will be to use extraterrestrial instruments. Perhaps by the year 2000 telescopes will be operating on the far side of the Moon, where they will be unaffected by light from Earth.

There are several reasons why no optical telescopes larger than 5 or 6 metres aperture have been constructed. Because of light pollution the magnitude limit which a telescope can reach does not increase as fast as the aperture, while the construction costs increase as the third or fourth power. So a 10-metre telescope will go perhaps

2 magnitudes fainter than a 5-metre telescope and cost at least five times as much. This partly explains why several 4-metre telescopes have recently been constructed; they are much cheaper than 5-metre instruments and perform nearly as well. Furthermore the development of better photographic plates, image intensifiers and electronic recording devices has meant that the light is now recorded with much better efficiency and has therefore lessened the need for even larger collecting areas.

2.3 Radio telescopes

The radio waves from space are collected either by large arrays of antennae or else by single dish instruments that are similar to optical reflector telescopes. Because radio waves are very much longer than light waves, the apertures of radio telescopes have to be large in order to achieve a reasonable resolving power. Since the earliest days of radio astronomy, its practitioners have built instruments of ever increasing resolution. An optical telescope with an aperture of 5 cm has the same theoretical resolution as a radio telescope about 5 km across operating at 6-cm wavelength. Since it is impossible to construct single reflectors with apertures of several kilometres, a technique known as interferometry has been developed by radio astronomers for high resolution work.

In an interferometer the signals from two small telescopes are brought together and analysed either electronically or by computer as if they were the signals received at two small elements of a giant instrument. A complete picture can be built up by changing the separation of the two telescopes, and by using the Earth's rotation to change the angle of the interferometer baseline relative to a celestial source. By moving aerials and observing as the Earth rotates it is possible to simulate the response of all possible pairs of elements in a much larger telescope (Figure 9). In order to speed up the process of aperture synthesis it is now usual to use several basic elements: there are eight in the 5-km telescope at Cambridge, England, ten in the Westerbork instrument (Netherlands) and twenty-seven are proposed for the Very Large Array in the U.S.A. It is usual to have a small on-line computer to control the telescope and perform most of the data reduction.

The largest dish-style telescope is the 300-metre 'hole-in-the-ground' telescope constructed in a natural crater at Arecibo in Puerto

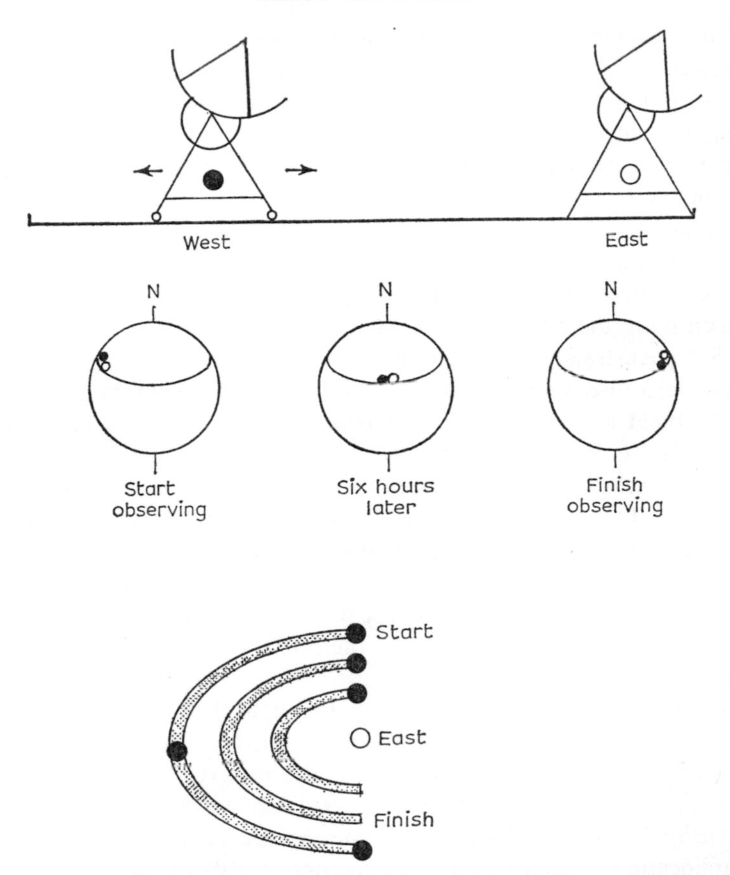

Fig. 9. In an aperture synthesis telescope one aerial (or antenna), here called West is movable. During a 12-hour observing period the separation is not altered, but the rotation of the Earth causes one dish apparently to rotate about the other. At the end of one 12-hour session an elliptical element of half of a giant telescope has been synthesized. On the following day a new separation is used and a further element swept out

Rico. It is operated by Cornell University and has made important surveys of faint radio sources. A disadvantage of the Arecibo instrument is that it can only see a limited part of the sky, close to the zenith. At the U.S.A. National Radio Astronomy Observatory in West Virginia is a 100-metre radio telescope which is steerable in elevation. In Europe there is a 100-metre fully steerable telescope at Effelsberg near to Bonn (Plate 2), and also the famous 80-metre dish at Jodrell Bank, which was the world's largest for more than a decade.

An interesting development of the early 1970s was that of very-long-baseline interferometry, in which telescopes on different continents can be linked to provide angular resolutions of a thousandth of an arc second. This is done by collecting the signals at the two separate stations on magnetic tapes, along with extremely accurate time signatures from an atomic clock. The tapes and clocks are then brought together for computer analysis, the time pips being used to synchronize them both exactly. By means of such links between the U.S.A., Australia, Sweden, U.K. and U.S.S.R. very fine structures have been discovered in some extragalactic radio sources.

To build a telescope or interferometer element with a parabolic dish reflector can be an expensive business. Consequently the early interferometers often consisted of arrays using parabolic cylinders as elements. These are less versatile in operation because they can only be steered in elevation. However, this does not matter in the kind of survey research which was used to produce the 3C and 4C catalogues at Cambridge, for example, because the entire sky has to be observed, and it is obviously sensible to use the Earth's rotation for sweeping the scan in one co-ordinate.

Radio astronomers do not look through their telescopes. The antennae collect a white noise from the cosmos which just sounds like a hiss if it is played through an audio receiver. The intensity of this hiss at different points in the sky is a measure of the radio intensity. Since the celestial radio sources are extremely weak great amplification is needed, often by factors of 10^{12} or so. To achieve this, special low-noise receivers have been developed, some of which use a maser as a sensitive high-gain amplifier. Typically, the background noise of the sky and from the telescope itself is far more intense than the radio source signal, and so techniques have evolved for extracting the signals from high noise levels.

Usually the radio intensity at a particular point will excite a signal of a certain voltage at the output of the telescope. The relationship between output voltage and the intensity of the celestial source is derived from calibration observations and the known properties of the telescope. The voltages were once used to activate pen recorders, but now it is more usual to feed the signal into digital counters or a computer. A common way of displaying the information is as a contour map, which shows the intensity rather like a topographical map shows the height of a hill (Plate 6). Newer techniques of visual display, which produce something analogous to an optical picture, are under development (Plate 20).

2.4 X-ray telescopes

Earth's atmosphere is opaque to cosmic X-radiation, and so observations of the X-ray emission from stars and galaxies have to be conducted from high-altitude rockets or from satellites orbiting Earth. A typical rocket flight allows only about one minute of observing time; consequently X-ray astronomy advanced fairly slowly until the launch in 1970 of Uhuru, the first satellite devoted to X-ray studies. This instrument produced a great deal of fascinating new information on binary X-ray stars and the X-ray emission from galaxies and clusters of galaxies. The positional accuracy of X-ray instruments is not yet as good as in radio astronomy, but it is expected to improve dramatically with the next generation of space observatories.

2.5 Spectroscopy

An important task of astronomy is to measure the intensity of radiation as a function of wavelength. In optical astronomy this is accomplished by dispersing the light into its constituent wavelengths with a diffraction grating, which produces results that are similar but superior to a prism's. The spectrum produced by the grating is then recorded either photographically or photoelectrically. Optical spectroscopy is used in investigations of the spectral lines of emission and absorption, because the intensities of the optically visible lines can be used to derive several important properties such as temperature, pressure and abundances of different elements.

In radio, infrared and X-ray astronomy it is more common to record the intensity at a few selected wavelengths and then to construct the spectrum by drawing a smooth curve through the points. The shape of the spectrum can be used to decide whether a particular source is radiating thermally or by some other mechanism, and it is also necessary for calculating the energy required to account for the emission. The spectra of radio sources are usually derived from observations made with several telescopes, as any one instrument has its best performance over a fairly small wavelength range.

A great many of the findings described in this book have been the result of optical and radio spectroscopy. It is by measuring the apparent wavelength of selected optical lines that we find the velocities of galaxies and then deduce how far away they are. Studies of the emission lines are used in deriving the conditions within exploding galaxies, radio galaxies and galactic nuclei. The mysterious quasars were first identified as something unusual because they have such baffling optical spectra. Radio astronomical measurements of spectra have helped to yield vital information on the staggering energy requirements of radio galaxies and quasars, as well as giving a picture of the actual emission process. Spectroscopy is of the greatest importance to explorations of the universe.

2.6 Sky surveys

An important part of fundamental research is the production of sky surveys. Messier and the Herschels made great surveys to catalogue the nebulae, and in ancient times Ptolemy drew up the Almagest which detailed the brighter stars. In astrophysics it is becoming increasingly important to know the properties of particular objects across a whole range of frequencies; the radio astronomer needs to correlate radio sources with their optical counterparts, and the X-ray astronomer may wish to know if an X-ray emitter is visible at infra-red wavelengths, for example. To assist these endeavours surveys of the sky are made within particular frequency bands.

Perhaps the most well-known survey is the Palomar Observatory Sky Survey, which was made with a 1.2-m Schmidt camera. On plates covering a field $6.6° \times 6.6°$ the whole sky visible from Palomar was photographed, recording stars down to about the 20th magnitude. Copies of this survey may be found in most of the world's major observatories. The survey is being extended to cover all the southern hemisphere using Schmidt telescopes in Australia (Plate 3) and Chile. From the Palomar survey a great deal of data on galaxies and clusters of galaxies has resulted. For example, the important catalogues of galaxies compiled by Fritz Zwicky and his collaborators were based on the Palomar photographs. In the Palomar survey red-sensitive and blue-sensitive plates were exposed for each field. By comparing red and blue plates it is possible to get a rough indication of the colour of objects.

The radio counterparts of the Palomar survey are the various catalogues compiled by the world's radio observatories. Among these the most famous is the Third Cambridge, or 3C, catalogue which lists over 400 objects. Radio source catalogues listing thousands of objects have subsequently been produced. There is no radio survey quite like the Palomar optical atlas, however, because the very deep radio surveys have tended to concentrate on a limited area of the sky. X-ray astronomers can use the 3U, or Third Uhuru catalogue, which lists a couple of hundred objects. It is similar in scope and extent to the 3C radio catalogue, which was produced about fifteen years earlier.

Radio and X-ray sources are often matched to optical components in the following manner: the celestial co-ordinates of thousands of bright stars are stored on a computer disc or magnetic tape. The position of the radio source is fed into the computer, which then searches its tape for about a dozen stars in the vicinity of the radio source. Next the positions of the stars and of the radio source are etched onto transparent film by a graph plotter attached to the computer. This map of the stellar and radio positions is made at exactly the same scale as the Palomar Atlas prints. The astronomer then positions the transparent overlay on the print so that the star positions on the film coincide precisely with those on the print. Now the area in the immediate vicinity of the radio or X-ray emitter is pinpointed so it can be scrutinized for a plausible optical counterpart. It is in just this way that all the radio galaxies and quasars have first been selected for further spectroscopic investigation.

3 · Distances

3.1 Distances to the stars

Fundamental to any complete understanding of heavenly bodies is the ability to determine how far away they are. Only when we know the distance to a particular galaxy can we calculate how much energy it radiates or how large are its component parts. These data are essential if we are to construct theories of galactic evolution. Also measurement of the distances to the galaxies is a vital step towards unravelling the large-scale structure of the universe.

Mankind has always thought the stars to be more remote than the sun, moon and planets, but it was not until 1838 that the distance to a star, 61 Cygni, was actually measured by the German astronomer Bessel. Until the distance scale had been established for stars it was not possible to fix the location of the extragalactic nebulae. Even today our surveying tools for galaxies ultimately rely on the distances to stars being known, and for this reason we first look at the determination of stellar distances.

Bessel used the method of trigonometrical parallax to find the distance to 61 Cygni. The technique exploits the fact that our Earth revolves around the sun on an orbit of radius about 150 million km. The length of this radius is called the Astronomical Unit. As the Earth goes on its annual motion, the direction of a nearby star will appear to alter by a small angle 2p, relative to faraway stars, on

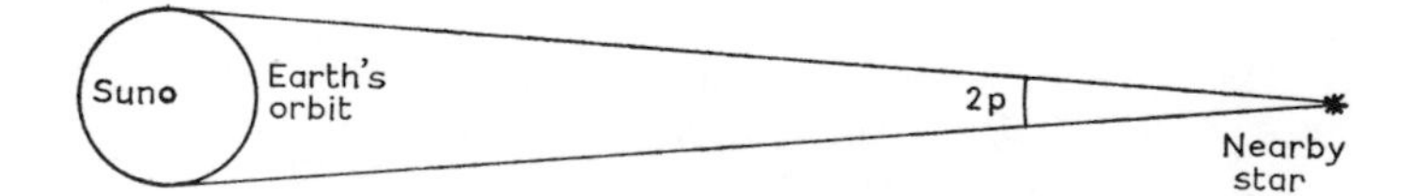

Fig. 10. During a year the apparent position of a local star changes by an angle, 2p, as Earth goes on its orbit round the Sun

account of the Earth's movement (Figure 10). In reality the change in angle through the year is so small that it can be measured only for a few thousand of the nearest stars. One-half of the displacement angle, p, is called the *parallax* of the star. Of course, the background stars will also shift their positions very slightly. However, modern computers can quickly derive, by statistical techniques, the absolute parallax that would be obtained if the background were infinitely far away. Parallax angles then are an indication of the proximity of stars; the angles decrease as the stars studied become more distant.

Stellar parallax is expressed as an angle, in practice always less than one second of arc (abbreviated as 1 arc sec = 1/3600 of a degree). 61 Cygni has a parallax of 0.293 arc sec, close to Bessel's estimate of 0.3 arc sec. The nearest star so far discovered is Proxima Centauri which has a parallax of 0.765 arc sec. These angles are far smaller than can be discerned with the naked eye; even with modern telescopes and computers the determination of parallaxes to within 0.003 arc sec is exacting and tedious. In practice a series of photographic observations, spaced over many years or even decades is needed to achieve the desired accuracy.

A convenient unit of distance for professional work is the *parsec*, abbreviated pc. This is defined as that distance for which the parallax is 1 arc sec. It equals 30.86 million, million km. Expressed in parsecs, the distance of a star with parallax angle p is 1/p pc. In popular writing the unit of distance most commonly encountered is the light year, the distance traversed in one year by electromagnetic radiation and equal to 9.46 million, million km. There are 3.262 light years in a parsec. Proxima Centauri is 4.3 light years away, and Sirius, the brightest star in the sky is 8.7 light years away. When dealing with distances across galaxies the kiloparsec (kpc) is frequently used: it is 1000 pc or 3262 light years. Our own sun is about 10 kpc from the centre of the Galaxy. Once we start striding out the other galaxies it becomes necessary to speak in terms of megaparsecs (Mpc), equal to 1,000,000 pc.

Beyond 300 pc the method of parallaxes breaks down completely. At this distance the parallax is comparable to the errors of the measuring process, so that no useful information can be extracted. In fact, even to penetrate to 30 pc by this classical surveying technique requires a high degree of care and precision, which has only been achieved for a few thousand stars. If we had only this triangulation method at hand, then the task of mapping out the structure of our

Galaxy and the universe would be utterly impossible. Fortunately, the trigonometrical survey of our near neighbours in space provides us with a direct calibration for a host of other measuring rods.

3.2 The moving cluster method

A further basic method of finding stellar distances is from measurements of the motion of stars in a nearby cluster such as the Hyades. This method gives distance directly and does not depend on the fact that the Earth orbits the sun. We could therefore use the moving cluster method to start calibrating the distance scale even if we lived on a stationary planet.

All the stars that are members of a given cluster move through space along essentially parallel tracks as the cluster travels through the Galaxy. If the space motion vectors of these stars are measured and then plotted on a map of the sky, all the vectors converge to a common point on the sky towards which the family of stars is travelling. The distance to the cluster is found by comparing the apparent angular motion of its members with the true space velocity. To find the latter the radial velocity (the component of motion directed along the line-of-sight to Earth) is found spectroscopically by measuring Doppler shifts of the cluster's stars. Now the tangential velocity has to be calculated, by multiplying the radial velocity by tan θ, where θ is the angular distance between the star's actual position and the point of convergence. Armed with this tangential velocity the distance is derived directly from the angular velocity, or proper motion. This moving cluster method is valuable for the Hyades and Ursa Major clusters, because these are both beyond the reach of direct triangulation, and they contain examples of the 'standard-candle' stars that are used to extend the distance scale across the local group of galaxies.

3.3 Candles instead of rulers

A whole series of distance-measuring techniques in astronomy is based on a comparison of the brightness of a star as we see it, with the amount of energy which we deduce that it is actually radiating. The apparent brightness of a star depends on the inverse square of its distance: put it three times as far away and the energy received by a particular telescope will fall to one-ninth of its original value;

if it is 10 times as far away the energy is diluted 100 times, and so on.

To put this argument into an astronomical context a definition of the term magnitude is needed. Unfortunately, star brightnesses were arranged into magnitudes in antiquity, long before there was any question of objective measurement, and consequently a cumbersome system arose. What happened was that in 130 B.C. Hipparchus drew up a catalogue in which the brightest stars were marked first class, or magnitude 1, and the faintest visible to the naked eye as magnitude 6. The remaining stars of intermediate brightness were assigned magnitudes 2, 3, 4 and 5 in order of diminishing brightness. In the nineteenth century the system was modified, first by use of decimal divisions (2.4, 3.8, etc.) and second by adoption of a fixed brightness ratio between one whole magnitude and the next. The ratio in brightness encompassed by a change of 5 magnitudes was fixed at 100 times. Therefore a 1-mag star is apparently 100 times brighter than a 6-mag star. A difference of 1 magnitude represents a ratio $100^{1/5} = 2.512$ in brightness.

Hipparchus and the early astronomers were only interested in the apparent magnitudes of stars—a measure of the quantity of starlight received at Earth after it has spread out during its journey across space. Since the stars are all at different distances, for physical discussions the concept of *absolute magnitude* must be used. By definition this is the apparent magnitude a star would have if it were placed 10 pc away. Thus, absolute magnitude, because it is tied down to a fixed distance, is directly related to the real brightness or energy output of the star. Let this real brightness, termed the luminosity, be L and the absolute magnitude be M. In the metric system the two are related as follows

$$M = 2.5 \log_{10} (L/3.0 \times 10^{28})$$

where L is measured in watts.

Suppose now that a star is a distance l away, and it has an apparent magnitude m; the absolute magnitude M is then given by

$$m - M = 5 \log_{10} (l/10)$$

when l is in parsec. From this relation it can be seen that the quantity $m - M$ is directly related to distance l and so is referred to as the *distance modulus*. Equivalently we can write the last equation as

$$\log_{10} l = 0.2(m - M) + 1.$$

Now we have the clue to a powerful method of distance measurement. We can measure m for a star or galaxy by means of a photometer; alternatively, the size of its image on a photographic plate can be measured and converted to an apparent magnitude. If we also have some method of deducing M, the absolute magnitude, then we can calculate the distance *l*. Fortunately, in a number of special cases, it is possible to estimate M in a manner which is independent of distance. In short, there exist 'standard candles' in the sky; by comparing real and apparent candlepower we can deduce distances.

Stars do differ greatly in their luminosities, ranging from 100,000 times more powerful than the sun, to feeble objects radiating 10,000 times less than the sun. Not many stars have special labels telling us their absolute magnitudes. A further difficulty is that the apparent magnitude can be reduced by obscuring gas and dust in space between us and the star. If the light is seriously dimmed in transmission because of interstellar dust, then the straightforward formulae above will not apply. Even so the general principles are still of assistance. And not only does the method work for stars, but as we probe deeper into space we can apply it to the apparent and absolute magnitudes of clusters of stars or even entire galaxies.

3.4 Standard candles in the star clouds of Magellan

The key that would ultimately crack the problem of long-distance surveying and open up galactic explorations was found at the Harvard College Observatory (Cambridge, Massachusetts) in the early part of this century from studies of the Magellanic Clouds, the two larger star systems beyond the Milky Way, which are visible from the southern hemisphere. Astronomically these dwarf galaxies are important because they are our nearest galactic neighbours, being about 160,000 light years away. They contain relatively large numbers of the giant variable stars known as Cepheid variables.

Harvard College set up an observatory in the mountains of Peru in order to survey the Small Magellanic Cloud. As has been mentioned, in 1908 Miss Henrietta Leavitt, a research assistant working on the Peruvian data, drew attention to an important property of the Cepheids. She found that the stars with the longest cycle of variation were also those of greatest brightness. By pursuing this correlation she produced unequivocal evidence for a relationship between period and apparent magnitude for Cepheids in the Magellanic

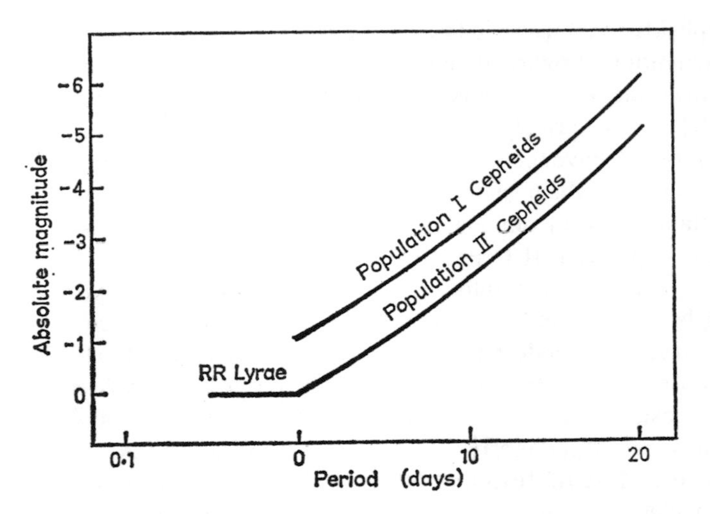

Fig. 11. The period of variability of a Cepheid or RR Lyrae is related to its absolute magnitude as shown here

Clouds. The period definitely increased for stars of brighter apparent magnitude (Figure 11).

The importance for distance measurements of the discovery of this relationship is that all stars in the Magellanic Clouds are at essentially the same distance, given that the depth of the Clouds is only a small fraction of the intervening distance. Therefore, Miss Leavitt's work indicated that the absolute magnitude correlated with the period for Cepheids, since apparent magnitudes in the Clouds are convertible to absolute magnitudes by addition of the same constant for all stars, namely the Cloud distance modulus m — M. In order to calibrate absolutely the period-luminosity law for Cepheids, it is necessary to find independently the distance of prototypes in our Galaxy. Harlow Shapley derived the law from trigonometrical measures of the distances of nearby Cepheids, and thereby transformed the Leavitt relation to one between period and absolute luminosity.

In the last twenty years the precise form of the period-luminosity relation has been derived as a result of much effort in the U.S.A. on the part of Walter Baade, Halton Arp, Robert Kraft and others, who showed that the early calibrations were seriously in error. For example, Baade's work showed that there were two types of Cepheid, following slightly different period-luminosity relations. Population I

4

Cepheids (stars with a relatively high proportion of metals) are some 2 magnitudes brighter than Population II Cepheids (metal-poor stars) of the same period. The Population II variables include a type known as RR Lyrae variables. These latter have periods of less than one day, and they all have the same absolute magnitude, $M_V = 0.5 \pm 0.2$ mag, a property which makes them invaluable as standard candles. Distances within our Galaxy had been determined, prior to 1950, from Population II Cepheids, for which the period-luminosity relation had been correctly established. But in other galaxies the Cepheids observed are often Population I objects and their luminosities had been underestimated, leading to distances that were five times too small. In 1961 Robert Kraft published a revised period-luminosity curve for Cepheids in our Galaxy. He has also redetermined period-luminosity relations for Cepheids in the Magellanic Clouds, M31, IC 1613 and NGC 6822. This important research on Cepheids in nearby galaxies showed the relation to be sharply defined, so that reasonable accuracy in distance measurement using Cepheids is now thought to be possible.

3.5 Galactic distances from variable and exploding stars

The method of distance measurement using Cepheids is now firmly established as a powerful weapon in the astronomer's armoury. It requires observers to determine the period P and mean *apparent magnitude* $\langle m \rangle$ of Cepheids in an external galaxy. The mean *absolute magnitude* $\langle M \rangle$ for a Cepheid of period P days is given by

$$\langle M \rangle = -1.41 - 2.36 \log_{10} P.$$

Having measured P in days (it will be between 1.2 and 40 days), $\langle M \rangle$ can be calculated. The distance in parsecs to the galaxy containing the star is then obtained from

$$\log_{10} d = 0.2(\langle m \rangle - \langle M \rangle - \delta m) + 1.$$

Here $\langle m \rangle$ is the mean apparent magnitude obtained photographically or photoelectrically. An additional quantity δm has been inserted to compensate for the absorption of starlight by interstellar gas and dust. In a particular case it will usually be estimated from the amount of reddening present in the light from the stars.

For the Large and Small Magellanic Clouds recent results yield distance moduli (i.e. values for $\langle m \rangle - \langle M \rangle$) of 18.6 and 19.0 mag respectively. The corresponding distances are 52,000 and 63,000 pc. In the Andromeda Nebula M31, Edwin Hubble, Walter Baade and Sergei Gaposchkin between them found several hundred Cepheid stars. These objects provide an accurate fix of 0.67 Mpc for the distance to M31. Cepheids show that other galaxies in the Andromeda group, namely M31 and M33, are about 0.7 Mpc distant. Galaxies in the Ursa Major group—including M51, M81 and M101—are close to the limits of the Cepheid method, since they are about 3.2 Mpc away. Beyond 4 Mpc it is impossible to use Cepheids; the distance modulus then reaches 28 mag, beyond which the apparent magnitudes of Cepheids are too faint for useful research.

Apart from the regularly variable RR Lyrae and Cepheid stars, other stars are sometimes used as intergalactic milestones. Out to 10 Mpc novae can be found and observed at maximum brightness. Stars of this type can increase in magnitude by 15 mag in only a day or so.

Different types of novae are recognized, and from studies of them in M31 and the Galaxy the maximum absolute magnitude can be fixed at −6 to −10 mag, depending on type. Halton Arp's survey of the novae in M31 enabled him to derive a relationship between the light curve and maximum brightness. From this link the light curves of new novae can be used to deduce absolute magnitude, and hence to act as distance indicators. In practice they are useful in confirming distances derived from other indicators, particularly the Cepheids. Attempts have also been made to use supernovae—fiery explosions marking the deaths of massive stars. However, these cataclysms exhibit quite a range in magnitude, and have hardly reached the status of an intergalactic measuring rod.

3.6 Hydrogen clouds as candles

Beyond about 4 Mpc other indicators of distance modulus must be used. In the spiral and irregular galaxies, which are rich in gas, huge clouds of ionized hydrogen exist. In these concentrations, called the H II regions, the hydrogen atoms have been stripped of their solitary electron, usually by the ultraviolet radiation from hot stars. The H II regions show a marked correlation in the sense that the diameters of the largest H II regions in a particular galaxy are a function of the absolute magnitude of the galaxy. From studies of galaxies within 4 Mpc, for which H II region diameters and galactic magnitudes can be found independently, the form of this diameter-luminosity relationship has been determined. Measurement of the apparent magnitude of a galaxy at unknown distance, along with the angular diameters of the H II regions within it, then permits the determination of distance. This method holds for those Sc, Sd, Sm and Ir galaxies that contain H II regions up to a distance of 60 Mpc.

3.7 The far universe

Beyond 20 Mpc or so it is common to use a variety of rough indicators of galactic distance. Within 60 Mpc, for example, there is a large enough sample of type Sc I galaxies for us to establish that their absolute luminosities do not vary markedly; in fact the difference in luminosity will generally be less than 50 per cent for two Sc galaxies chosen at random. This fact gives us an opportunity to use the energy output of an entire galaxy as a standard candle. In terms of absolute magnitude, M = −21.2 mag for the Sc galaxies within 60 Mpc. So if we assume this value also applies to Sc galaxies further away a distance can be assigned from their apparent magnitudes. To penetrate further still—out to 1000 Mpc or more perhaps—Allan Sandage introduced, in a series of papers published in 1972, the method of brightest cluster galaxies. In rich clusters of galaxies many of the members are elliptical type, and the brightest galaxy in the cluster is also E-type. The brightest ellipticals in clusters appear to be a remarkably homogeneous set of objects with a mean absolute magnitude M = −21.7 mag. Sandage has demonstrated that the variation in absolute magnitude among the brightest galaxies in the great clusters is 0.32 mag. Such a small scatter increases confidence that the apparent magnitudes of the brightest cluster members can be used as a distance indicator.

When probing up to 1000 Mpc or so, by magnitude-distance relations, the question of intergalactic absorption clearly assumes importance. After all, prior to the present century, astronomers were tricked by interstellar absorption, which led to incorrect estimates of the distances to faraway stars. In turn this led to wrong models for the Galaxy. How can we be so sure that intergalactic dust is not cutting down the light from remote galaxies? Fortunately, the methods of estimating distance do interlock to some extent and therefore there are opportunities for cross-checking which have not led to serious discrepancies. Also, Sandage has shown that, for brightest members of clusters, the apparent angular diameter correlates very closely with apparent magnitude. It would be difficult to account for this result if intergalactic absorption were severe.

3.8 The Hubble diagram

In 1929, one of the golden moments for extragalactic astronomy and cosmology occurred when Edwin Hubble established a relationship between the velocity of a galaxy and the distance to it. Slipher, of the Lowell Observatory, had measured the radial velocities (i.e. velocity with respect to the sun) of about forty galaxies. He did this by taking spectra and then determining the apparent shift in spectral lines from their rest wavelength. If a line is shifted by an amount $\delta\lambda$ from the rest wavelength λ, because the source of emission (i.e. the galaxy) is moving with speed v, then

$$v = (\delta\lambda/\lambda)c$$

where c is the speed of light. This change in the wavelength of the lines due to velocity is known as the Doppler effect. The relation between velocity v and shift $\delta\lambda$ given above holds unless v is a significant fraction of c.

By 1929 Hubble had the distance to about twenty-five galaxies available, and in that year he reached a monumental conclusion: the galaxies are racing away from us with a velocity of recession that increases with distance from us. He gave a linear relation between velocity v and distance d:

$$v = Hd.$$

Here, H is a constant with dimensions (velocity/distance), and it is expressed in (km/sec)/Mpc.

Hubble gave H = 500 km/sec/Mpc from his studies of Cepheids and the brightest stars in nearby galaxies. The calibration of the luminosity-distance relation for these standard candles was corrected in the 1950s by Walter Baade, who was the first to recognize the existence of the two distinct stellar populations. Each of these had different luminosity-distance relations, as explained above.

In 1968 Gustav Tammann and Allan Sandage made an important breakthrough in establishing the distance scale by isolating and studying Cepheids in the galaxy NGC 2403. They set its distance as

3.25 ± 0.20 Mpc, and found 65 km/sec/Mpc as the average of velocity/distance for galaxies in the same grouping as NGC 2403. Note, however, that this value for a single grouping cannot be described as a value for H, since we may be moving relative to the galaxies studied with a velocity over and above the systematic effect.

Allan Sandage pushed the business of accurately surveying the galaxies further out a little later in 1968. Within the great Virgo cloud of galaxies lies the giant spherical galaxy M87. Individual globular clusters of stars can be studied in M87, and so Sandage suggested matching the brightest clusters in M87 with the brightest clusters in our own Galaxy; he was relying on the assumption that both have the same absolute magnitude. This procedure places M87 at 14.8 Mpc from the Galaxy. For the Virgo cluster a mean value of 77 km/sec/Mpc was arrived at for velocity/distance. However, the mean velocity of recession for galaxies in the Virgo cluster is 1136 km/sec, and at this level there is still no guarantee that random motions within the cluster of galaxies are not making a significant contribution. Allan Sandage and Sidney van den Bergh have emphasized independently that the distances to the Virgo galaxies will be in error if the proposed comparison of clusters of stars in M87 and local galaxies is invalid. George Abell has argued that M87 may not be a standard candle because the Virgo cluster as a whole displays irregularity and is relatively loosely structured. All these considerations for M87 show how difficult it is to lay down precise values for extragalactic distances beyond 4 Mpc or so.

Sufficient uncertainty exists in this field of astronomy that it may be most unwise to rely on any one distance indicator to determine the constant H. Sidney van den Bergh published in 1970 a mean value based on the results of several investigators who had worked independently. He gave

$$H = 95 \pm 15 \text{ km/sec/Mpc}.$$

Even so, the best we can safely conclude is that H is between 50 and 100 km/sec/Mpc.

The Hubble law provides the key to unlock the secret of galactic distances beyond 10 Mpc. Once a redshift $z = \delta\lambda/\lambda$ is established spectroscopically for a galaxy, a provisional distance estimate can be obtained from

$$d = \frac{zc}{H},$$

where c is the velocity of light. As an illustration, consider a galaxy with a redshift 0.06, take c = 300,000 km/sec and assume that H = 100 km/sec/Mpc. Then we obtain

$$d = \frac{0.06 \times 300{,}000}{100}$$
$$= 180 \text{ Mpc.}$$

Therefore the galaxy is 180 Mpc distant if its redshift of 0.06 is due solely to the systematic recession of the galaxies.

In recent years the Hubble law has come under heavy attack on several fronts. Despite the criticisms that have been made, it is still thought to be reliable provided the galaxies considered are normal spiral and elliptical galaxies. For the brightest galaxies in clusters Sandage has given evidence that the Hubble law holds fast to redshifts of 0.46. The distance scale is therefore extended to the immense distance of 1500 Mpc and perhaps even further, by application of the relation to galaxies of known redshifts.

3.9 Are extragalactic distances reliable?

Our measuring rods of the universe seem to be constructed like some gigantic pyramid (Figure 12, Table 2). On each level the calibration of the rod depends on all the levels below holding up! Starting at the most basic level, we rely on radar pulses bounced from the planets to determine the size of the solar system. Then classical triangulation with Earth's orbit round the sun as baseline is used to reach the nearest stars. More distant stars are pinpointed by the moving cluster method or spectroscopic techniques. By this means we find out how far away rich clusters of stars in the Galaxy are. These clusters contain variable stars, particularly Cepheids and RR Lyrae variables. The fact that they are in clusters of known distance enables us to set the magnitude-period relation for such stars on an absolute base, thus giving an invaluable source of standard candles. Variables make up our stepping-stones to the nearest galaxies. Having mapped out the local extragalactic neighbourhood, the apparent magnitudes of brightest stars, sizes of H II regions and properties of star clusters can be used to reach the nearest giant clusters of galaxies. By this stage sufficient data has been accumulated to tie down the Hubble relation, which can then be applied to penetrate to the galaxies of intermediate distance. The distances to the

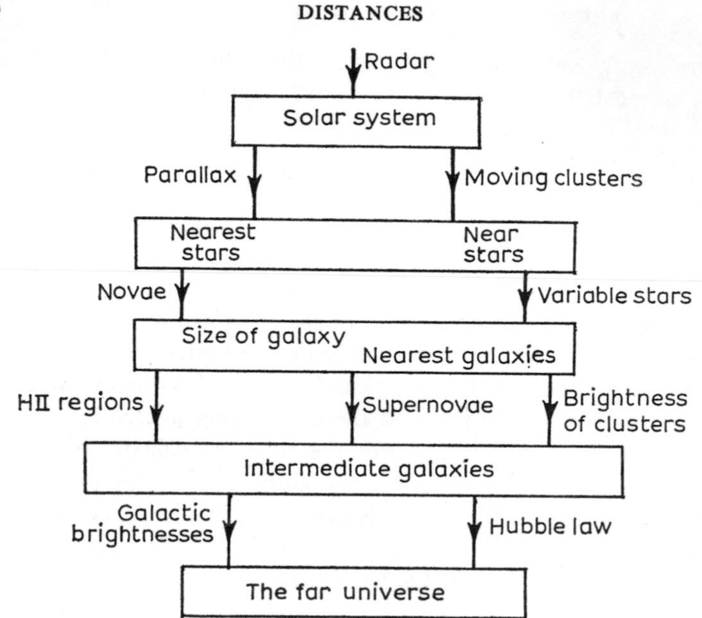

Fig. 12. The methods of determining astronomical distances

most remote galaxies can only be guessed at. This is usually done by assuming that they are the brightest members of remote clusters, and taking an absolute magnitude $M = -21.5$ mag for the brightest cluster galaxies. Since galaxies down to apparent magnitude $m = 24$ can now be routinely photographed, it is possible that galaxies with distance moduli $m - M = 45$ have been photographed. They would be 10,000 Mpc away if all our assumptions are justified; it takes light 32.6 billion years to travel such a colossal distance.

Although every layer of the pyramid relies on the strength of all the layers below, it is fortunate that plenty of cross-ties exist to strengthen the structure. For many of the nearer objects we can use several independent methods of distance estimation to check our values. It is unlikely that the distance of well-studied objects such as the Magellanic Clouds or M31 can be in error by much over 10 per cent. Between 1 Mpc and 10 Mpc the errors should not amount to more than 25 per cent for most 'normal' objects. In the regions where the Hubble law is conventionally assumed we may generally be within a factor of two of the right distance. Our cross-checks interlock to a sufficient extent that statistically there is not much latitude for

TABLE 2

Distance determination

Method	*Application*	*Normal effective range*	*Typical uncertainty*
Celestial mechanics	solar system	0.001 pc	extremely small
Radar	solar system	0.001 pc	extremely small
Trigonometrical parallaxes	nearby stars	0–100 pc	5–50 per cent
Moving clusters	nearby open clusters	40–500 pc	5–20 per cent
Spectroscopic methods	clusters in our Galaxy	up to 15 kpc	20–50 per cent
Cepheids and RR Lyrae stars	globular clusters and local galaxies	2 kpc–4 Mpc	10–30 per cent
Novae, red giants, supergiants	local galaxies	2 kpc–4 Mpc	25–50 per cent
Sizes of H II regions	nearby galaxy clusters	1 Mpc–10 Mpc	25–50 per cent
Supernovae	nearby galaxy clusters	100 kpc–100 Mpc	25–50 per cent
Integrated magnitudes of globular clusters	nearby galaxy clusters	up to 10 Mpc	25–50 per cent
Brightest galaxies in great clusters	very distant clusters	10–1000 Mpc	? 50 per cent
Redshift	very distant clusters	10–1000 Mpc	? 50 per cent

more deviation than that. Beyond 1000 Mpc it is true that we are doing no more than estimating the distance, but unfortunately we have no choice. New instruments for studying very faint objects, such as image tube cameras, may one day enable us to find a standard candle or well-calibrated ruler for these remote objects.

4 · Explorations of normal galaxies

4.1 Stellar sociology

A casual glance at the sky reveals the familiar patterns called the constellations. These groupings are not physical associations of stars in space, but the result of projecting the real distribution of stars onto the flat sky. Photographs of the sky do, however, reveal that many stars are organized in close-packed communities. Objects like our sun are essentially single stars without any strong relationship to the other nearby stars. However, many stars occur in binary and multiple systems, and still others are gathered into star clusters of various sizes, ages and types.

There are two principal types of star family: the *galactic* or *open clusters*, which have a loose structure and contain perhaps a few hundred stars, and the compact spherical *globular clusters* which are dense affiliations of as many as a million stars. In addition to clusters there are also extensive groupings of specific types of star, and these are termed associations.

Some of the nearest galactic clusters can actually be seen without a telescope. The bright stars in Ursa Major and the Hyades are easy to see, and the Pleiades make a splendid sight in the winter skies of the northern hemisphere. In this latter group a person with average sight can see up to eight stars, and low-powered binoculars will reveal several dozen. In southern skies M7 in Scorpius is a brilliant galactic cluster. So far, several hundred galactic clusters have been catalogued, and it is clear that many thousands must exist in the Milky Way, although most of these are so far away that they are lost in the rich starry background or obscured by dust and cannot be perceived. They are confined to the plane of the Milky Way. In nearby external galaxies the open clusters are resolvable, and they tend to be concentrated in the spiral arms.

There are sound reasons for believing that all the stars within a given galactic cluster condensed from the interstellar gas at roughly the same time. In the Pleiades, which is a particularly youthful family, the vestiges of the interstellar gas still festoon the cluster. The coincidence of star ages in a cluster leads to a powerful method for finding its distance from a determination of its distance modulus (m — M). When the colour-magnitude diagram of a distant cluster is plotted, the stars define a *main sequence*, and also perhaps a *giant branch*. At this stage of the investigation only the *apparent* magnitudes of the stars are used. But the plot can also be made for nearby clusters of known distance, such as the Hyades, using *absolute* magni-

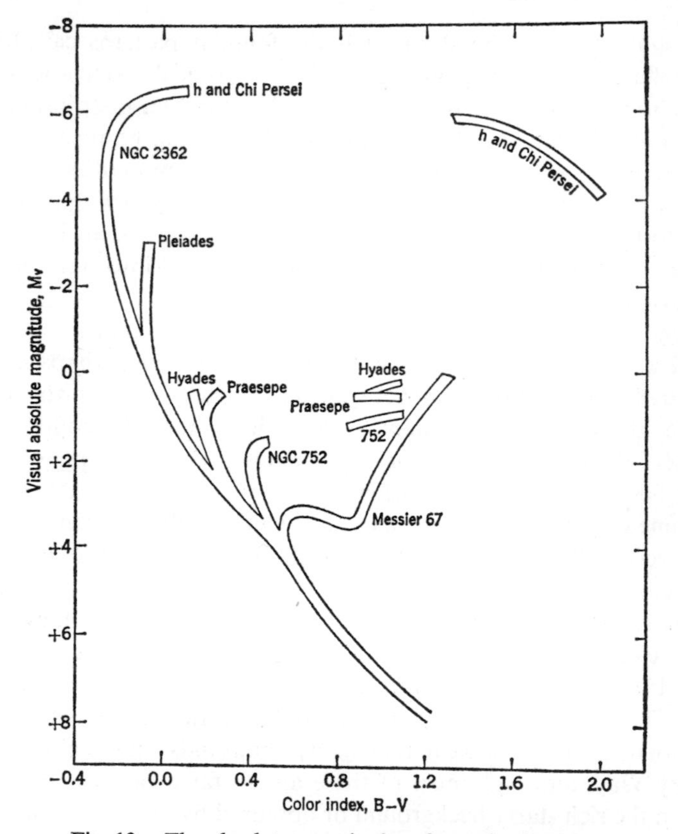

Fig. 13. The absolute magnitudes of stars in clusters can be found by matching the main sequence of a cluster to that for stars at a known distance. Here several clusters are shown thus superposed

tudes. The vertical displacement in magnitude (m – M) that brings the main sequence of the faraway cluster (plotted with apparent magnitude) into coincidence with the main sequence on the graph for a cluster of known distance (plotted with absolute magnitude) is a measure of the distance modulus (Figure 13). This technique of 'main sequence matching' was pioneered at the Lick Observatory in the 1930s by R. J. Trumpler, and it was his research on this topic which led on to the important discovery of interstellar obscuration.

Globular clusters are like enormous balls of stars, and the most magnificent among them, ω Centauri and 47 Tucanae, are visible as 4th mag fuzzy stars in the southern hemisphere. The brightest in the northern hemisphere is M13 in Hercules. Through a telescope some thousands of stars can be seen, mostly fainter than 11th mag, against a luminous background of a myriad of unresolved stars. Globular clusters are nothing like as numerous as galactic clusters. About 100 are known, and many of them were first catalogued by Messier or the Herschels. The globulars are much less concentrated to the plane of the Milky Way, and are in fact distributed in a great spherical halo enveloping the Milky Way. Harlow Shapley determined the distances to globular clusters using the Cepheid relationship discovered by Miss Leavitt. In 1918 this gave him the first hint of the colossal size of the galactic system, when he showed that the halo distribution spanned some 100 kpc.

Although the globular clusters are ancient, stable systems, this is not the case for galactic clusters. Not only are the latter less populated to start with, but they are also moving on orbits in the galactic plane, where encounters with other stars or even clusters may be quite frequent. This means that the cluster will gradually dissolve and so lose its strong identity. In the late 1940s Ambartsumian discovered *stellar associations*, which are extensive distributions of loosely associated stars, and which may represent the last phase in the break-up of star clusters like the Hyades. Many have now been recognized, the most conspicuous being the vast Scorpio-Centaurus association. Neither the galactic clusters nor the associations are such permanent features of galactic structure as the globular clusters.

4.2 Stellar spectra and composition

The spectra of the great majority of stars can be conveniently divided into a number of spectral classes designated by a letter of the alphabet. Each class is defined by the prominence in the spectrum of the lines of selected elements. Although the boundary between classes is not sharp it is usually possible to assign a spectral classification unambiguously. When the classification technique was originally evolved at the Harvard College Observatory the spectral types were indicated by the letters A, B, C, . . . Later it was discovered that some groups were superfluous, and that the surface temperature played a critical role in determining the type. Thus a more sensible order, based on physical principles, O, B, A, F, G, K, M, R, N, S, was eventually used. The main features of stars in each type are listed in Table 3, along with the surface temperatures.

One of the most significant diagrams ever constructed in astrophysics is the Hertzsprung-Russell diagram, often abbreviated to the HR diagram. In this the absolute magnitude M_V of a star is plotted against spectral type. Since the spectral type correlates with temperature it is essentially a graph of energy emitted from the star (absolute magnitude) against its temperature (spectral type). With the advent of UBV photometry it became more common to plot the directly observable quantity (B–V) instead of spectral type or temperature. Essentially the same results are obtained, and the graph is then known as a colour-magnitude diagram.

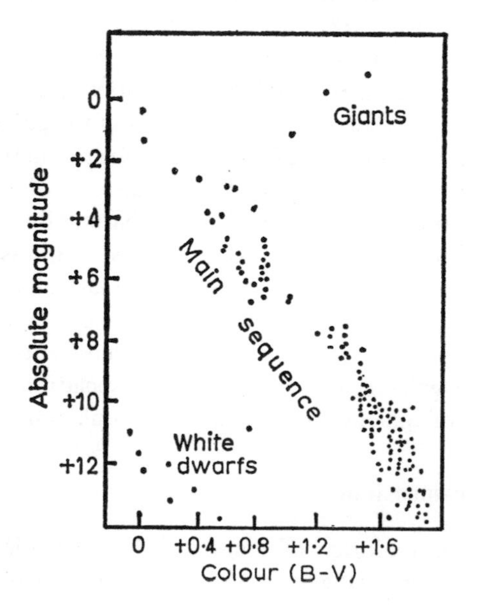

Fig. 14.
The HR diagram for the nearest stars

TABLE 3

The classification of stars by spectral type

Spectral type	*Characteristics*	*Effective temperature*	*Typical examples*
O	Hot stars with absorption due to ionized helium; strong ultraviolet continuum.	38,000–32,000 K	ξ Pup, λ Cep α CMa, τ CMa
B	Absorption lines of neutral helium dominate; hydrogen developing.	12,500–30,000 K	β CMa, δ Ori α Paw, λ Cen β Per
A	Hydrogen lines at maximum; ionized calcium increasing.	8,200–10,800 K	α CMa, α Pic α PsA, β Tri
F	Ionized calcium stronger, hydrogen weakening, metals developing.	6,200– 7,200 K	β Vir, α For δ Gem, α Car
G	Solar-type spectra, metals strong, hydrogen weak.	5,500– 6,000 K	α Aur, β Hya κ Gem, α Ret
K	Metallic lines dominate, molecular bands developing.	4,200– 5,200 K	α Tau, α Bov
M	Very red stars with strong absorption bands of titantium oxide	2,700– 3,900 K	α Ori, α Hya α Cru, ω Cyg

When the HR or colour-magnitude diagram is constructed for the nearby stars of known distance a diagram shown schematically in Figure 14 is obtained. Note that stars do not scatter around to fill the whole graph; instead they order themselves into four major groupings. The band containing the overwhelming majority is called the *main sequence* and the other groups are called *giants*, *supergiants* and *white dwarfs*; these names arise because their radii are either large or small compared to stars on the main sequence. It can be seen that the main sequence encompasses luminous (M_V high), blue (B–V small) stars at the top left-hand end, through to dim (M_V low), red (B–V greater than +1 mag) stars at the other. The blue stars are the hottest and the red stars are the coolest. When whole galaxies are considered, it is generally the distribution and properties of the hottest and most luminous objects that determine the overall appearance of the galaxy.

The usefulness of the HR diagram lies in the fact that stellar evolution makes a great difference to a star's position on the diagram. Objects on the main sequence, about 90 per cent of the observable nearby stars, are burning hydrogen to produce helium. During this phase they remain in more or less the same place, and the luminosity is proportional to a high power of the mass (L is proportional to M^4 for the more luminous stars). The conversion of hydrogen into helium leads to enrichment of the stellar interior, and at some stage a rearrangement of the star's internal structure becomes necessary. The object moves off the main sequence and becomes both brighter and redder, reaching the red-giant phase of stellar evolution. More massive stars make this transition sooner than low-mass stars.

Although the HR diagram shown in Figure 14 is for the nearby stars, for which distance is known and hence for which the absolute magnitude can be calculated, valuable results can still be obtained even when the distance is unknown, provided the observations are restricted to star clusters. Cluster dimensions are tiny compared with their distance from the sun. Consequently, we can construct a cluster HR diagram using just the apparent magnitude, or alternatively the magnitude in the V band, for the vertical axis. HR diagrams for clusters possess the main sequence and giant branches, but the spread in colour and magnitude is smaller than in the plot for nearby stars. This is because the objects in a given cluster form a more nearly homogeneous group in terms of age and initial chemical composition, with the mass being the major variable property amongst the individual stars.

There is a striking difference between the HR diagrams for globular and open clusters (see Figure 13). Globular clusters contain the oldest stars in the Galaxy; consequently all the stars that once populated the bright end of the main sequence have exhausted their hydrogen fuel and moved over to become red giants, so the globular clusters have a small main sequence that turns off and joins the giant branch. From the top of the giant zone there is a horizontal branch, within which the RR Lyrae variables are found. For the younger open clusters there is a considerable variation in the position of the main sequence turn-off point; the lowest turn-offs, corresponding to the oldest open clusters, just about match up with the turn-off for globular clusters. Studies of the HR diagrams for clusters within the Milky Way and in external galaxies have been extremely useful for plotting features of the evolution of galaxies, particularly the evolution of the stellar component.

In passing, a few remarks will be made about special types of star, because they will occasionally be mentioned in a galactic context. The white dwarfs are a very interesting group. They are the relics of stars that have exhausted all supplies of nuclear energy and then collapsed under gravity to a stable configuration where the pressure exerted by electrons is able to hold back the gravitational squeeze. White dwarfs are so dense that a matchbox of material would weigh a ton. Their intrinsic luminosities are very low, and it is only possible to observe them directly within a few tens of parsecs of the sun. In galactic studies one important feature of white dwarfs is that a large population of them contribute very little to the overall luminosity of a galaxy, but they may form a significant fraction of the mass, and probably do so in the case of elliptical galaxies.

Many groups of stars are variable; the light output changes with time, and these variations are due to a specific combination of physical conditions. Among the variable stars that exhibit regular fluctuations we have seen that the Cepheids and RR Lyraes are of outstanding interest to our pursuit of the structure of the universe.

Occasionally a star may undergo a great and sudden increase in luminosity and shine forth briefly as a nova or supernova. Explosive mass loss accompanies a rise in luminosity by many orders of magnitude. For novae the explosion is not totally devastating: some stars recover and suffer the process more than once. But for a supernova the explosion probably ruptures the star completely. Only four have been seen in our Galaxy in the last thousand years, not one since the invention of the telescope. Supernovae can rival the light output of a whole galaxy, and they are important because they can be observed even in quite distant galaxies. About a dozen are observed each year in external galaxies. The galaxy NGC 5253 is considerably fascinating because three supernovae have now been seen in it within this century, the last in 1972.

4.3 Galactic morphology

Galaxies differ greatly in size and brightness. Colossal systems such as the Milky Way and the Andromeda Nebula are many thousands of times brighter and more voluminous than the midget satellite galaxies that accompany them (Chapter 4). And there are many giant elliptical galaxies that are even larger than our Galaxy (Plate 4).

As we have already mentioned in Chapter 1, the system of galactic classification introduced by Hubble in 1925 is the one that is still in widest use. The Hubble classification is based purely on appearances or morphological consideration. Three main classes are recognized: spirals and barred spirals (S and SB), elliptical (E) and irregulars (I). Among the spiral galaxies three stages of development are distinguished; these are denoted Sa, Sb, and Sc (or SBa, SBb and SBc for a barred spiral). The relative size of the central regions decreases from Sa to Sc, and the relative strength of the spiral arms increases from Sa to Sc. In contrast to the spirals, elliptical galaxies possess no easily distinguishable internal structure, having a smooth progression from an elliptical boundary to a bright centre. The boundary has indefinite edges, so that an elliptical galaxy will appear larger in size on photographs going to fainter limiting magnitudes, which pick up the dim peripheries. They differ in their ellipticity, varying from the type E0 with its circular outline, through to E7, in which the ratio of the lengths of the major to minor axis is 3:1.

Among professional astronomers increasingly sophisticated classification schemes have been drawn up. Hubble himself added a type S0, to represent the lenticular galaxies. These are superficially like ellipticals but they have a central nucleus and also dark clouds on their peripheries, typical of spiral galaxies. Additional spiral stages Sd and Sm for specifying the more disorganized structures have been added. In later chapters a whole bestiary of strange galaxies will be introduced: Seyfert galaxies, radio galaxies, compact galaxies and quasi-stellar objects. For the present chapter we set the task of exploring only the ordinary run-of-the-universe specimens.

The relative numbers of galaxies of each type are of some interest. In lists drawn up on the basis of apparent brightness, such as the *Shapley-Ames* or *New General Catalogue* (NGC), spiral galaxies will be found to be in the majority. For the thousand brightest in the *Shapley-Ames Catalogue*, Harlow Shapley listed the proportions as 75 per cent spiral, 20 per cent elliptical and 5 per cent irregular. Even in modern times it is still the case that of all the galaxies investigated with any thoroughness about three-quarters are spirals. However, these proportions are unrepresentative of galactic populations as a whole because we have taken a bias towards the more luminous objects by using catalogues based on apparent brightness. It happens that galaxies of above average luminosity are more likely to be spirals.

If instead we make a census of all the galaxies in a given volume of the universe, paying no regard to brilliance, then large numbers of faint ellipticals and disorganized irregulars have to be included, and the correct percentages turn out to be 30 per cent spiral, 60 per cent elliptical and 10 per cent irregular, according to Shapley. Most of the photographs reproduced in popular astronomy books are of spirals. This is because they display a variety of structures, unlike the ellipticals, and also because the relatively strong blue light from the arms accentuates the structure recorded by the blue-sensitive emulsions normally used.

In a way spiral galaxies have personality. With some practice it is easy for one to match the catalogue numbers of all the brighter spirals in the NGC with a selection of unlabelled photographs, but it is almost impossible to do the same for ellipticals.

Although the elliptical galaxies are categorized on the basis of ellipticity, the degree of flattening observed in a given case is obviously determined as much by projection effects as by profound astrophysics. A thin rod viewed end-on looks circular, but it is not spherical in shape. Thus the E0 galaxies may be true spherical star families, or more likely, spheroidal systems viewed along the axis of rotation. The E7 systems are not sidereal cigars as many writers suggest; a cigar-shaped volume of stars is unstable against its internal gravitational forces and becomes ripped to pieces by conflict within.

Rather, the E7's are flattish wheels of stars viewed edge-on. In the intermediate classes, E1 to E6, tilt and true oblatenesses are mingled to an extent that cannot be determined without precise measurements of the velocity fields of the individual stars. One general statement can be made, however: an elliptical galaxy is at least as flat as it appears on a photograph!

The oblate shapes of the elliptical galaxies are easily explained. It is only necessary to accept that the galaxies possess angular momentum and hence they rotate. The flattening is then due to interplay of the centrifugal force associated with rotation and the self-gravitation of the stars within the galaxy.

Now that radio and optical telescopes are penetrating ever deeper into the universe it becomes necessary to classify minute smudges on photographic plates. It is often impossible to distinguish E7, S0 and edge-on spirals for these very distant systems.

4.4 Galactic luminosities

Most of our information about galaxies comes from measuring and analysing the light from their constituent parts. The distribution of brightness in a particular object is determined by photographic or photoelectric photometry, which gives information on the star density at different points in the galaxy. Care is needed in interpreting measurements of course, because variations in the predominant type of star, thickness of the galaxy, and internal absorption will all affect the measured luminosity per unit area.

Elliptical galaxies, because of their regular structure, have been the subject of many photometric studies. The isophotes, or contour lines joining points of equal luminosity, of elliptical galaxies are generally a set of nested ellipses of similar shape. Different elliptical galaxies appear to follow the same luminosity distribution, apart from scaling factors, as if they are all constructed according to a common scheme. For lenticular galaxies photometric studies have disclosed three features of galactic anatomy, namely the nucleus, the lens-shaped disc and an enveloping halo of stars.

Not surprisingly, spirals display the greatest variety in terms of luminosity distribution or 'personality'. For ordinary spirals the average brightness at different radii from the nucleus decreases exponentially. But the picture is complicated by the presence of bright emission knots within the arms and nuclear regions. As we have already noted the arms are not as conspicuous as most photographs suggest. When careful measurements are made of the total amount of light it is seen that the regions between the curvaceous arms are not starved of sidereal matter; rather, they are occupied by a population of dwarf stars. Pictures of galaxies made by image-tube techniques give a much more faithful account of the distribution of light; strangely, such pictures often look soft and fuzzy because the blue supergiant stars are not exaggerated as they are on photographs taken with blue-sensitive emulsions.

For nearby galaxies the total flux of luminous energy, or the absolute magnitude, is found in the following manner. First the integrated, or total, apparent magnitude is found, either by measuring out the individual isophotes and summing up over the area of the galaxy, or by recording the magnitude through a series of diaphragms of increasing radius and then by graphically extending the result to a defined limit. From these measured magnitudes a 'corrected' apparent magnitude is deduced. 'Corrected for what?' one may ask. Well, if the galaxy is near the plane of the Milky Way, interstellar dust will have dimmed its light before it reached our photometer. Also, if it has an appreciable redshift there are effects associated with the shifting of the bulk of the light to longer wavelengths, where the photometer may have a different sensitivity. A further very important point, and one on which astronomers are by no means agreed, is the diameter out to which the measurements should be taken. The giant galaxy M87 can be followed out across 0.5 degrees of arc with highly sensitive instruments. At what distance should a halt be called for purposes of defining a magnitude? Sandage, among others, has suggested that the data collection should proceed out to an agreed value of surface brightness (i.e. magnitude per square second of arc). For faint galaxies these difficulties are not so important as the image size is smaller and it is not always essential to have extremely accurate measurements. But if galactic magnitudes are used in cosmological investigations, as Sandage has done, then very painstaking effort is essential; all the magnitudes must be measured in the same manner. Once total apparent magnitude m is derived, the distance modulus $m - M$ is computed from the distance, and thus M the absolute magnitude is deduced. As we mentioned in Chapter 3 it happens that for distant galaxies we may prefer to 'guesstimate' M and deduce the distance if we cannot find the latter independently.

An important index of galactic populations, much used in censuses of clusters and different galactic types, is the *luminosity function.* Generally this function is presented as a graph showing the numbers of galaxies N(M) in successive absolute magnitude intervals. The luminosity function for *complete samples* shows that there are many dwarf galaxies; these are difficult to detect on account of their intrinsically feeble performance. For example the luminosity function of all the objects in the *New General Catalogue* shows a different state of affairs because the cataloguer selected them on the basis of *apparent magnitude.* Consequently, the bright giants are over-represented and the faint dwarfs poorly represented. The NGC catalogue is an example of an *incomplete sample*: because the dim objects are disfranchised, the luminosity function of its members tells us little about the general properties of galaxies. The question of completeness is of vital importance when a corpus of observational data is used for general theoretical or cosmological work. If an important component has been completely missed, and this fault is not taken into account, the deductions may be worthless.

The absolute magnitudes of galaxies range from a maximum of about —22.5 down to a minimum of —8.5 mag. The apparent magnitudes of detectable galaxies run from about 4 for M33 (visible to the naked eye) down to the present technology barrier at about 25 mag.

4.5 Weighing the galaxies

During the few thousand years of civilization and the systematic quest for knowledge, man has progressed from making measurements with the simple balance, to the abilities to measure minute sub-atomic particles that live for infinitesimal fractions of a second and to determine the masses of mighty galaxies, far out into the universe, that have existed much longer than the Earth. There are a handful of observational methods for finding the masses of the galaxies; some of them are more trustworthy than others and the various techniques will not necessarily yield the same answer for a given object.

The most reliable method, and the one to which most observational effort has been devoted by optical and radio astronomers, is based on the internal rotation of spiral galaxies. On a cosmic time scale the great wheeling galaxies are rotating quite rapidly. In a typical giant spiral the stars make one circuit in about 100 million years; the sun takes 250 million years to make one orbit round the Milky Way. Individual stars and gas clouds are moving at speeds of up to 300 km/sec. Even so, there is no hope whatever of determining the rotation law for a given galaxy simply by taking a pair of photographs spaced in time by a few decades. Instead, the spectroscope is used as a speedometer for logging the velocities of parts of distant galaxies. For nearby galaxies it is preferable to obtain the emission-line spectrum of many emission nebulae at different distances from the galactic centre. The spectra have good sharp lines from which it is easy to determine the relative velocities of the gas nebulae by

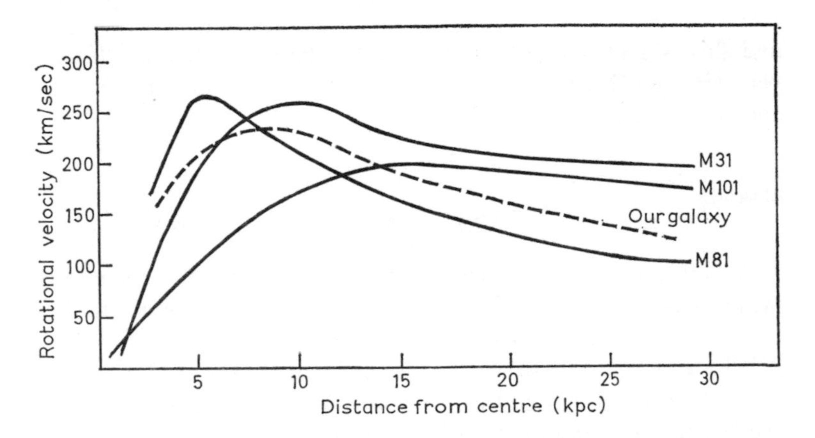

Fig. 15. Variation of rotational velocity with distance from centre for the Galaxy, M31, M81, and M101

assuming that differences in the wavelengths of spectral lines are due to relative motions within the galaxy. A graph called a rotation curve can then be constructed; it shows the velocity of rotation of the galaxy as a function of distance from its centre (Figure 15). This method requires substantial time on a large telescope because the speeds at several points in the galaxy must be sampled. A large number of these rotation studies were made by Geoffrey and Margaret Burbidge and Kevin Prendergast in the 1960s. Similar observations can now be made by radio astronomers, mapping the precise frequencies of 21-cm line radiation from hydrogen in the distant galaxies.

The gravitational field within a galaxy determines the orbits of its constituent parts. This field is related to the distribution of mass, which in turn tells us how much mass there is in a galaxy. So the rotation curve is, in fact, determined by the mass of the galaxy and the way in which that mass is distributed. Generally theorists prefer to make a model of the mass distribution, and use this to derive mass estimates from the rotation curves.

A quick method of finding a mass is to measure the velocity v of some stars a distance r from the centre of the galaxy. One then assumes that they are moving on a simple orbit like a planet round the sun; the centrifugal force per unit mass is

$$f = v^2/r$$

and this is assumed to be balanced by the gravitational attraction per unit mass GM/r^2, where G is the universal gravitational constant and M the galactic mass Thus

$$GM/r^2 = v^2/r$$

and so

$$M = v^2r/G.$$

As a simple illustration we can consider the motion of the sun in the Milky Way. Taking $r = 10$ kpc, G as 6.67×10^{-11} Newtons $m^2\,kg^{-2}$ and v, the velocity of the sun as 250 km/sec, gives $M = 2.9 \times 10^{41}$ kg $= 1.5 \times 10^{11}$ solar masses as the mass of our Galaxy. This method gives a rough and ready value, but it is basically unreliable because the velocities are small at large r, and so they can be unduly influenced by random and systematic star motions. However, this method was used by early extragalactic researchers, who found that the giant galaxies near the Milky Way (e.g. M31 and M33) are many billions of times more massive than the sun.

The simple 'planetary' dynamics method outlined above is obviously no good for precise measurements. Kepler's laws of planetary motion assume that all the mass of the system is concentrated at the centre, which is patently untrue for galaxies. What we need is a method that accounts correctly for the fact that the matter in a galaxy is distributed across a large area. Mathematical models of galaxies must therefore be constructed inside computers. The dynamical behaviour of the stars within these galaxies can be studied as a function of total mass and the way it is distributed. Then we have to pick the model that most satisfactorily mimics the rotation curve derived observationally. Of course, to prevent the computers running riot and printing out a great range of masses for each galaxy, certain other conditions are used to constrain the models. The model distribution of mass must be consistent with the observed distribution of light, for example.

Results for the rotation-curve method show that galactic masses range from 10^{10} to 10^{11} times the sun for most bright spirals. The Andromeda galaxy is a true colossus: weighing in at 4×10^{11} solar masses, it is the most massive spiral known.

So far our discussion has centred on spiral galaxies. For the ellipticals and irregulars it is hard, or even impossible, to get worthwhile rotation curves. In the case of ellipticals one dodge that can be tried for getting a rough mass is just to analyse the widths of spectral lines in the absorption spectrum of the galaxy. This width says something about the velocity mix for stars in the galaxy, and that in turn can be linked to a mass. This approach has yielded 4×10^{9} solar masses for M32, one of the companions of M31.

A powerful method applicable to pairs of galaxies is a treatment analogous to that used for finding the masses of binary stars. Many galaxies are apparently double. Many of these galactic twins have speeds which strongly suggest mutual attraction and interaction, in which the pair is bound by gravitational forces. The work of Erik Holmberg and Thornton Page has accumulated a great deal of information on galaxy pairs. Although we do not know enough about the orbits of any given pair to enable an accurate mass determination, the data can be treated statistically to yield values for average masses and the way in which this varies for different types of galaxy. Thornton Page gave the mean masses of fourteen pairs of spirals and irregulars as 2×10^{10} solar masses and of thirteen pairs of ellipticals and lenticulars as 6×10^{11} solar masses.

Information on the masses of galaxies making up a discrete grouping or cluster can be extracted from observations of the positions and velocities of individual members. In order to do this it is assumed that the cluster is stable against dynamical disruption, and that there is no significant distribution of invisible matter. If these restrictions hold for a cluster then the virial theorem can be applied. This powerful theorem states that the sum of the time-averages of twice the kinetic energy and of the potential energy should be zero. As mass enters into the expressions for kinetic and potential energy the mass can be deduced from measured velocities and positions. In practice the masses derived for clusters in this way are often considerably higher than those worked out by other means, and the ratio of mass-to-light (in solar units) is anomalously large. These discrepancies indicate that clusters of galaxies may contain substantial amounts of 'missing matter', so the virial theorem method is discredited as a method of finding galaxy masses in clusters.

Once the luminosity and mass of a particular galaxy or cluster has been found it is a simple matter to derive the mass-luminosity ratio, or mass-to-light ratio, by dividing the mass in solar units by the luminosity in solar units. By definition $M/L = 1$ for the sun. Celestial bodies with M/L greater than this are less luminous than the sun, mass for mass. M/L is about 50 or so for the elliptical and lenticular galaxies but it can rise as high as 100. This indicates that the ellipticals contain substantial amounts of under-luminous matter, probably in the form of white dwarf stars, neutron stars or maybe even black holes. For spiral and irregular systems M/L is generally less than 10. The very luminous Seyfert galaxy NGC 1068 has $M/L = 2$, for M31 $M/L = 20$ and for M33 $M/L = 15$. In the case of loosely structured dwarf galaxies such as the Magellanic Clouds values of 5 or less are obtained. Note that all the values quoted exceed 1, the solar value: faint dwarf stars contribute very little to the light output of a galaxy but may make up most of the mass in fact. From the M/L values it is clear that the inconspicuous stars are more numerous in elliptical galaxies than in spirals or irregulars.

Values for M/L for clusters are very large indeed if the mass is taken from the virial theorem equation. For the great clusters in Virgo and Coma Berenices M/L exceeds 500. Such figures suggest that the underlying assumption of the applicability of the virial theorem may well be wrong, and particularly that there may be substantial amounts of invisible gas in the clusters.

4.6 Galactic colours

Different types of galaxies have different colours. The colour of a galaxy is found from photographic plates exposed behind filters or photoelectric measurements of its magnitude in the U, B and V wavelength bands (Table 4). The colour-index given by $(B - V)$, is the difference in magnitude in the B (blue) and V (yellow) bands. Another colour index is given by U (ultraviolet) minus V.

TABLE 4

The UBV filter system

Name	*Maximum response*	*Wavelength range of significant response*	*Equivalent width*
U	3500 Å	3000–4000 Å	690 Å
B	4350 Å	3600–5500 Å	960 Å
V	5550 Å	4800–6800 Å	900 Å

Galactic colouring correlates with galactic type as shown in Figure 16. The value of (B − V) decreases from +1 mag for elliptical galaxies, to 0.7 mag for Sc, and down to about 0.4 mag for irregulars. This range is smaller than that for stars, which runs from −0.3 to +1.6 mag. The smaller range is mainly a reflection of the preponderance of normal stars as the principal source of galactic light, irrespective of galactic type. A colour-colour diagram for galaxies is given in Figure 17. This shows that in a plot of (B − V) against (U − B) the galaxies occupy a broad swathe that is parallel to but not exactly identical to the sequence for normal stars. The colour-colour diagram reflects the steady progression from reddest (ellipticals) to bluest galaxies (irregulars). A given galactic type spreads

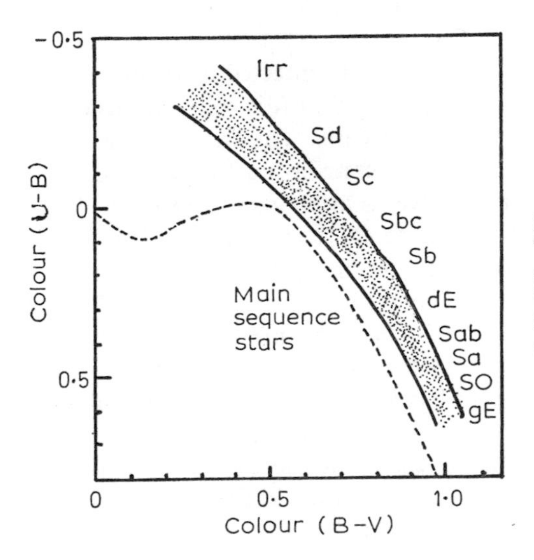

Fig. 16.
The correlation of galactic colour with type. The variation of colour is also shown for stars on the main sequence

inside the broad band of galactic colours because there is a spread in the average stellar contents between the galaxies of the same morphological type. If filters defining a narrower waveband than the UBV filters are used for photometric studies of galaxies it is possible to get a rough picture of the dominant stellar types in the galaxy.

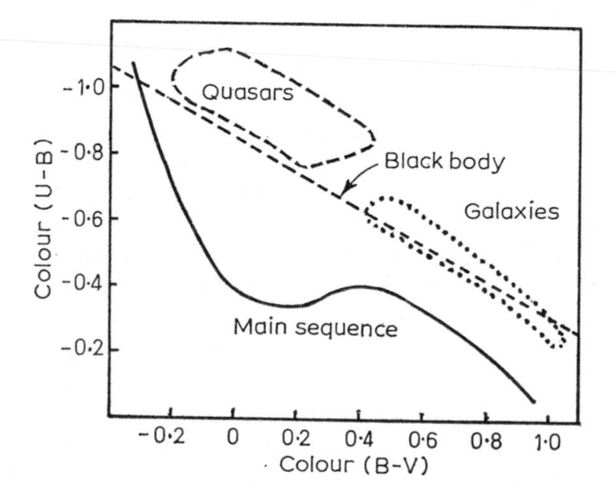

Fig. 17. Colour-colour diagram shows how colour may be used to discriminate galaxies, quasars, and main sequence stars

The intrinsic colour of a galaxy can be affected in several ways as the light journeys to our telescopes, so that it is necessary to correct the measured UBV colours in certain cases. The most obvious source of error is that due to reddening. Interstellar dust in the Milky Way and the distant galaxies themselves scatter light. Short wavelength light (blue) is more strongly scattered than long wavelength light (red), so that radiation travelling through the fine dust will get redder and redder, as the short-wave blue photons are preferentially scattered away from the direction of propagation. The reddening caused by our Galaxy can be determined by observing the colours of reddened stars of known spectral type and comparing the values of (B — V) with those of unreddened stars. A correction to the (B — V) colour for external galaxies in the same stellar field can then be worked out. Intrinsic reddening, due to dust in the distant galaxy, is harder to deal with. By comparing galaxies of the same type but with different angles of tilt to the line of sight Gerard de Vaucouleurs has derived average corrections for internal reddening and absorption.

4.7 Galactic spectra

Extragalactic spectroscopy can be said to date from 1864. In that year the wealthy, distinguished, English amateur astronomer William Huggins examined the great nebula in Andromeda. From the spectrum he concluded that the nebula must be composed of myriads of faint stars. Galactic spectra are normally governed by the stellar content, and spectra can be used to determine the dominant star type. Giant ellipticals, for example, have spectra indicative of a preponderance of red dwarf stars. In addition to the stellar component, the spectrum frequently shows absorption lines of ionized calcium atoms—so-called H and K lines—as well. These features are impressed by gas inside the galaxy and their apparent wavelengths are often used to calculate the redshifts of galaxies.

A small proportion of galaxies are sufficiently excited to show strong emission lines in their spectra. Included in this category are the Seyfert galaxies, which have the richest emission-line spectra. The most commonly observed lines are those of singly ionized oxygen, [O II], at 3726 and 3729 Å. Note that these lines are in the ultraviolet; if they are exceptionally strong they increase the magnitude in the U filter-band. Consequently, galaxies with strong emission lines do not fall onto the normal galactic sequence in the colour-colour diagram. With the advent of ultrafast automatic measuring systems for astronomy it will be possible to seek out unusual galaxies by programming a computer to search plates for galaxies with anomalous colours.

For decades, extragalactic spectroscopy was exceedingly difficult. Galaxies have low surface brightness, necessitating long exposure times. Only small dispersions could be used, so that the entire spectrum of a faint galaxy might spread across a plate with an area no more than a fingernail, with consequent loss of resolution. Fortunately the situation is rapidly changing in two important respects. First there has been the technology breakthrough. Far more sensitive instruments are now used to detect the light from faint galaxies. Image-tube intensifiers, for example, hold on to many more of the precious photons than the traditional photographic plate. In the most sophisticated systems, light from the galaxy can be dispersed through a grating and then analysed in real time by a small computer in the telescope dome. As soon as sufficient photons have been collected for the purpose in hand, the observer can switch to the next observation. Modern electronic and computing techniques are thus being used to uprate the world's greatest telescopes and so draw the beckoning galaxies into better view. The second breakthrough has been the enormous growth of interest in extragalactic affairs. Results from X-ray and radio astronomy especially have made astronomers thirst after much more information on the colours and spectra of galaxies. Analysis of the spectra from the emission-line galaxies is used to determine temperature, particle densities and energies within the excited gas causing the emission.

5 · Inside the galaxies

5.1 The stellar contents of galaxies

The notion that galaxies contain distinct populations of stars goes back to 1944, when Walter Baade introduced the concept of two populations, I and II, within our Galaxy and M31. His basic ideas have been of importance to all subsequent discussions of the structures and evolution of the galaxies. During the wartime blackout in 1944 Baade secured plates of M31 and its two satellite companions. By use of the 2.5-metre telescope at the Hale Observatories he was able to make long exposures in good seeing, and from his observations constructed the Hertzsprung-Russell diagram of stars within the nuclear bulge of M31 and inside the two companions. When this was done he could see that the composite HR diagrams resembled those of globular clusters in our Galaxy. The brightest stars in the regions investigated were red supergiants. When the outer regions of M31 were sampled a different story emerged, for here the stars yielded a composite HR diagram typical of galactic clusters, with the brightest stars being blue objects on the main sequence. Baade called the stars of galactic-cluster type *Population I* and those of globular-cluster type *Population II.* At first it was assumed that these two populations reflected a difference in age; although this is basically true, it is now evident that chemical composition, especially the presence of metals, plays a part in distinguishing the populations.

In 1951, Baade continued this work by sampling the two globular clusters NGC 6528 and NGC 6522, which are visible in a relatively unobscured zone south of the nucleus of our Galaxy. Measurements of 100 RR Lyrae variables in these two clusters gave a distance from the sun of about 9 kpc, in confirmation of their proximity to the nucleus. The HR diagrams strengthened Baade's view that galactic nuclei in spirals are mainly composed of ancient stars, like those in the globular clusters scattered through the halo of the Galaxy. Baade's original division was according to location; the nuclear and halo regions contain Population I stars and the disc comprises Population II stars.

Since Baade's pioneering work several modifications have been made. It is now clear that there is no sharp distinction between the populations, but more of a gradual merging. Furthermore theoretical advances in stellar evolution have shown that both age and composition are determining factors as far as the Population type is concerned. In Table 5 the current division of galactic matter into various Population I and Population II sub-groups is given, along with an indication of typical members, and a theoretical estimate of the ages of each group.

TABLE 5

Membership of the two basic stellar populations

Population I Objects	*Population II Objects*
open clusters	globular clusters
sun	weak-line stars
interstellar gas	galactic bulge
T-Tauri stars	novae
strong-line stars	RR Lyrae variables
giant stars	long-period variables
classical Cepheids	high-velocity stars
	metal-poor stars
	Population II Cepheids

An understanding of the interplay between stellar evolution, stellar populations, age and composition is essential for progress in interpreting observations of the galaxies. This is because only the gross properties of galaxies such as magnitude, colour, integrated spectrum and structure can be determined with any ease. These properties are influenced by the relative proportions of the populations and their positions within a galaxy. When the dominant star types and their locations are determined for many galaxies it is possible to consider that part of galactic evolution which is determined by the history of star birth, life and death.

5.2 Gas and dust

Optical, radio and infrared astronomers have shown that, in addition to stars, galaxies contain considerable amounts of gas and dust. The proportion of gas and dust compared to the total mass of a galaxy varies with the type, but it is generally highest for the irregulars and lowest for ellipticals. By far the most common element in the gas is hydrogen. In its atomic form this can be detected by means of the 21-cm radio line. This particular atomic transition is associated with the tiny difference between the two lowest energy states of hydrogen. To get from one state to the other the single electron in the hydrogen flips itself over, emitting a photon of 21-cm radiation if the transition is to the lowest possible state, and absorbing one if it is moving out of this state. It happens that neutral hydrogen is sufficiently common in our Galaxy and nearby galaxies to make the radio mapping of the location and velocity of the gas possible.

The neutral gas is mainly gathered into the spiral arms of the Milky Way, and in places it forms dense concentrations known as H I regions. The neutral hydrogen gas observed in the radio spectrum accounts for 3 per cent of the total mass of the Galaxy. Inside the dense H I regions there are probably abundant supplies of hydrogen molecules. These cannot be detected by radio emission, but if allowance is made for their almost certain presence, the fractional mass in the form of hydrogen comes out at approximately 6 per cent. A second very important constituent of the gas is helium, the second lightest element. On average there is one helium atom for every ten hydrogen atoms. Since the helium atom is almost four times more massive than the hydrogen atom, the proportion by mass is higher, and about 28 per cent of the mass of the galactic gas is due to helium.

In addition to the lightest elements there is a fair sprinkling of heavy substances, including nitrogen, oxygen, carbon, neon, magnesium and iron. Altogether these 'heavy' elements make up just under 2 per cent of the mass of interstellar matter. Their number contribution is much smaller because they are individually heavy. When the grand total mass of gas is calculated, it comes to something approaching 10 per cent of the total mass of the Galaxy. The remainder of the mass resides in the stars.

The gas in the Milky Way is not distributed uniformly. Mention has already been made of the fact that neutral hydrogen, which is detectable through the 21-cm line, tends to be found in the spiral

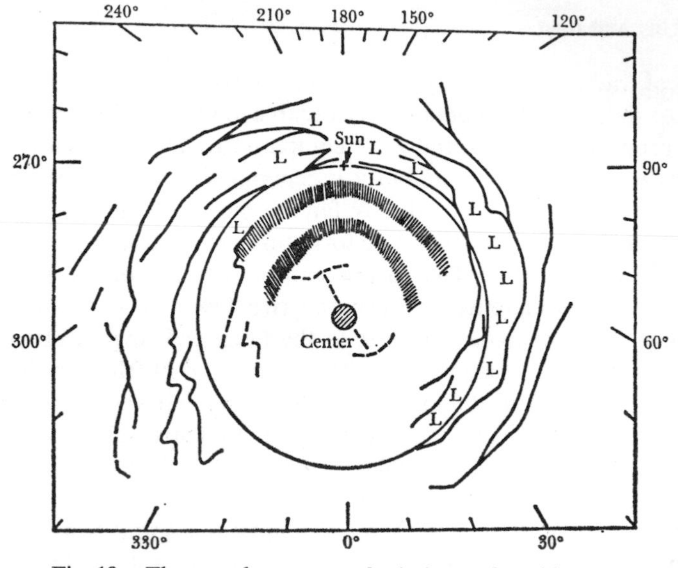

Fig. 18. The complex tracery of spiral arms found in our Galaxy by radio astronomers

arms. It is this very property that has enabled radio astronomers to map out the whirlpool pattern traced by the spiral arms of the Milky Way (Figure 18). This is done by determining the velocity, with respect to the sun, of gas concentrations seen in different parts of the sky. Small deviations from the rest wavelength of the 21-cm line are due to gas motions (the Doppler effect), and can be used to produce charts of the velocity of the gas as a function of the celestial co-ordinates. Often, several clouds can be seen, moving with different individual velocities, in the same part of the sky. Interpreting these charts requires knowledge of the motion of material in the Milky Way. This information has come from the study of star motions, which have revealed the gross features of our Galaxy's rotational behaviour. Assuming that the gas obeys approximately the same dynamical rules, it is possible to deduce from the measured velocities the unknown distances to the clumps of gas. When this is done for the neutral hydrogen its location in the Galaxy is mapped out and it is found that the hydrogen traces out a vortex of spiral arms. A given line of sight may cross several arms, moving with different velocities, and this is why more than one velocity can be observed in many parts of the Milky Way.

From the late 1960s onwards, a series of spectacular discoveries was made of organic and inorganic molecules in galactic gas clouds. Particularly noteworthy were studies of the huge concentrations at the galactic centre, the hub of the Milky Way, in which over two dozen distinct molecular species have been identified, including substances with five, six, seven and even nine atoms. The molecules found include formaldehyde, carbon monoxide, formic acid, acetyldehyde, cyanogen, ammonia, methyl alcohol and dimethyl ether. Some of the molecules are widely believed to be essential prerequisites to the natural synthesis of sugars, amino acids, nucleic acids and proteins, so the discoveries have given a great boost to studies of the extraterrestrial origin of life.

Among the visible manifestations of the interstellar gas are emission and reflection nebulae. These galactic nebulae are local concentrations of the interstellar gas, and those visible to the naked eye include the Trifid Nebula in northern skies and Eta Carinae Nebula in southern latitudes (Plate 5).

When an emission nebula, such as Orion, is examined with a spectroscope, it is found that the bulk of the light is concentrated into a few sharp emission lines, generally of ionized oxygen, neon, possibly iron and of neutral hydrogen. The fact that the heavy elements are ionized, that is to say have lost one or more of their electrons, shows that the gas is highly excited and has a high temperature. For many years astrophysicists were greatly puzzled by a line at 5376Å, which had been observed in several nebulae, but which could not be identified with any known atomic transition. They coined the word 'nebulium' to describe the mysterious atoms responsible for the unidentified line radiation. Eventually, physicists showed that ions of iron that had lost nine of their outer electrons were causing the radiation. So many electrons can only be stripped off if the gas temperature rises to tens of thousands of degrees. Inside a typical emission nebula the electron temperature is of order 10,000 K and the particle density is of order 10^8 per cubic metre. The excitation of the emission nebulae can usually be traced to the presence, inside the nebulae, of very hot and luminous young stars, mainly O stars. Almost certainly these exciting stars have themselves formed from gas in the nebula. Reflection nebulae occur when the interstellar medium is rendered visible by reflecting light from a nearby star.

When a hot star is embedded in a cloud of neutral hydrogen, the intense ultraviolet radiation from the star has a drastic effect on the hydrogen. Its outer electron absorbs sufficient energy to depart from the central proton. So the hydrogen atom is dissociated, or ionized, into its constituent proton and electron. Close to the hot star the hydrogen gas is transformed into a plasma of protons and electrons, which is termed an H II region. These objects emit thermal radiation which is detectable as a continuous spectrum of radio waves. Radio source surveys have led to the discovery of many H II regions in our Galaxy, some of which cannot be seen optically on account of obscuration. Our own galactic centre has a great concentration of giant H II regions. In external spiral galaxies the H II regions are often very conspicuous, both to optical and radio telescopes, and this is particularly true for the face-on spiral M33.

Dust in interstellar space is detected primarily through the blocking and obscuration of optical radiation. Photographs of spiral galaxies viewed edge-on show dense masses of dust in the galactic plane. The radio galaxies Cygnus A and Centaurus A, for example, possess gigantic swathes of dust that cut off the light from the central parts of the galaxy altogether. Across the band of the Milky Way it has long been noted that certain areas are relatively or totally devoid of stars. To see this on a clear, dark night the Milky Way should be viewed with the naked eye, and in both northern and southern hemispheres it will be seen that there are dark patches in the bright background. The Coalsack is especially conspicuous in southern skies for example. Composite photographs of the Milky Way show these dark zones very clearly. It might be supposed that the patches are genuine voids through the stars, through which one is looking towards the inky blackness of the extragalactic universe. For this hypothesis to be true, however, we would have to assume that empty tunnels through the stars in the Milky Way are sufficiently numerous for one tunnel to intersect our line of sight for each dark patch. It can easily be shown that the number of tunnels required to sustain the hypothesis is much too large. Hence, the voids must be due to obscuration. The most efficient way of shutting off the light from the star fields is by means of dust in interstellar space.

Further evidence for the existence of dust came from the discovery of interstellar reddening. Suppose stars of a given spectral type are observed at different distances. Then it is observed that as the distance increases the stars become redder in colour and their apparent magnitudes decrease more rapidly than increased distance effects can account for. These effects were first seriously investigated around the turn of the present century. The explanation of the reddening is that interstellar dust is scattering the light from the distant stars. It turns out that the probability of scattering depends on $1/\lambda^4$, where λ is the wavelength of the radiation. Now red light has a longer wavelength than blue light, so the value of $1/\lambda^4$ is smaller for red than blue light. This means that red-light photons are less likely to be deflected from the beam of starlight than blue-light photons. So, on its journey from the star to Earth the starlight progressively loses relatively more light from the blue end than from the red end of the spectrum. Consequently, the starlight is reddened by the interstellar dust, and the effect becomes increasingly noticeable at larger distances.

Interstellar dust haze also causes an overall dimming of the light from distant stars. Since the sun is located in the plane of the Milky Way this is an extremely serious matter, because over most of the plane it is not possible to see stars that are more than 2 or 3 kpc

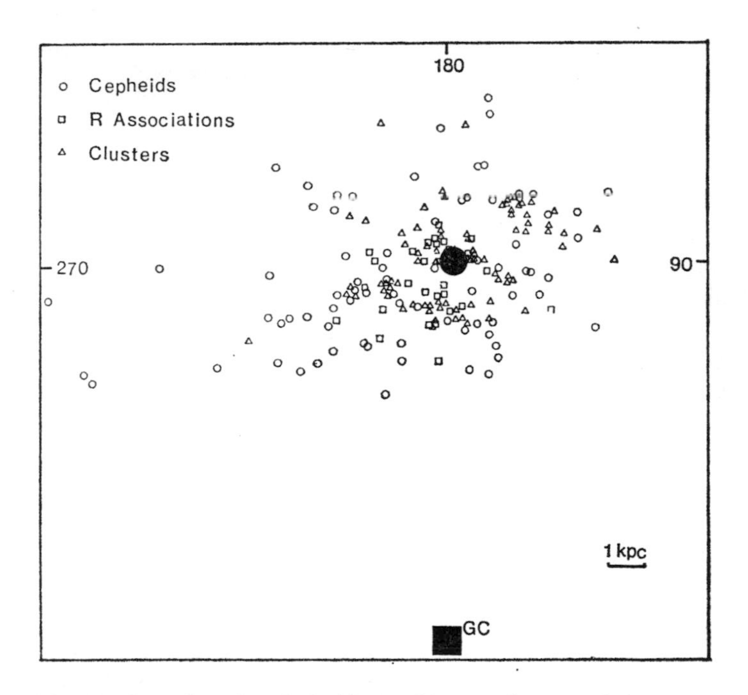

Fig. 19. Location of optical objects of known distance relative to the sun (filled circle) and galactic centre (filled square). It is just possible to discern segments of three spiral arms

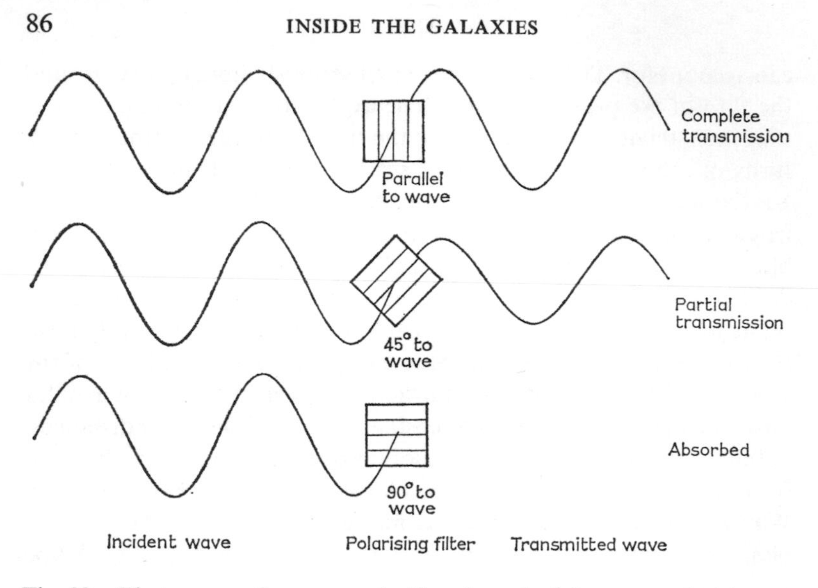

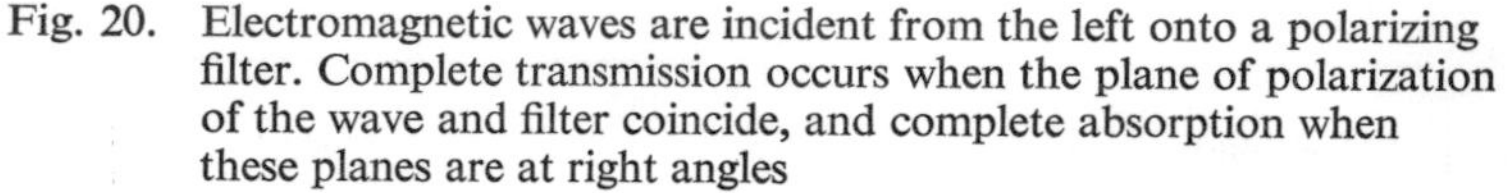
Fig. 20. Electromagnetic waves are incident from the left onto a polarizing filter. Complete transmission occurs when the plane of polarization of the wave and filter coincide, and complete absorption when these planes are at right angles

away (Figure 19). To be sure, there are occasional 'windows' through which we can peer deeper into the Galaxy, but these are neither large nor common. Spectacular dust structures can be seen in some H II regions, where they form long, dark lanes known as elephant trunk systems. Note that the famous Horsehead Nebula in Orion is in fact a cloud of dust which is silhouetted by bright backlighting from stars.

Information on the size and composition of the dust grains has come primarily by studying the dependence on wavelength of the absorption caused by the dust. The silicate grains, which are oxygen-rich, have average radii of 0.15 μm, the graphite grains which are composed of carbon are 0.05 μm in radius, and the iron grains are the smallest at 0.02 μm. These are just average values obtained from observations in several parts of the Milky Way. In order to account for the observed extinction of starlight, the amount of dust needed in the Galaxy is about 1 per cent or less, by mass, compared to the gas. It can readily be seen that interstellar dust accounts for 0·1 per cent or less of the mass of the Milky Way, since the gas itself contributes some 10 per cent and the dust is only one-hundredth of this.

A value of slightly less than 2 per cent was quoted above for the percentage by mass of heavy elements. In the interstellar medium dust grains are made of heavy elements, but they are responsible for under 1 per cent of the interstellar medium's mass. This means that heavy elements must be concealed in some unobservable form. These could take the form of a population of complex molecules that have not yet been discovered in the gas, or even of 'snowballs' in interstellar space.

Interstellar dust grains cause polarization in the light from stars. This effect can be demonstrated by rotating a polaroid film in front of a photometer that is monitoring the light from a star. If the starlight contains a component of radiation that is vibrating in only one plane (a component of linear polarization), a periodic fluctuation in the signal passing through the polaroid and onto the photometer is observed, because the polaroid blocks the linear component when it is vibrating at right angles to the principal plane of the polaroid film (Figure 20). By using a rotating polarizing filter in conjunction with a photometer it is possible to measure the percentage polarization and the angle of polarization for the stars. Polarization surveys have revealed large-scale systematic effects in the behaviour of linearly polarized light from the stars. These phenomena correlate with galactic position and are not intrinsic to the stars. What has happened is that cylindrical dust grains have been regimented by the galactic magnetic field so that their major axes are perpendicular to the field lines. As starlight traverses these aligned grains the radiation becomes polarized by scattering. By studying the run of polarization

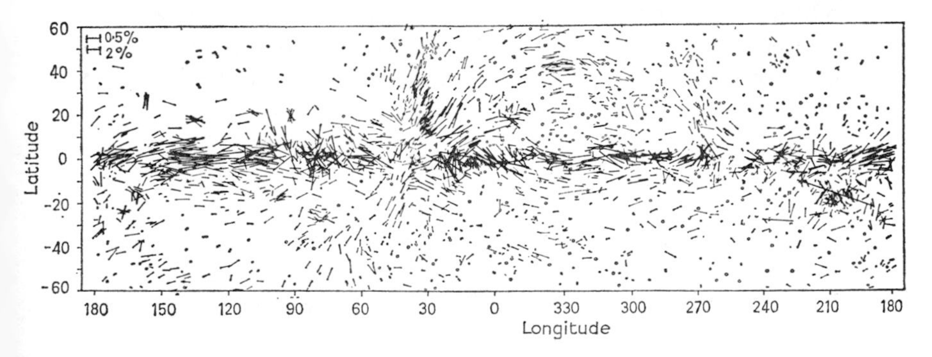

Fig. 21. The galactic magnetic field is revealed by the polarization vectors of stars

phenomena with galactic latitude and longitude it has become possible to learn something of the structure of the galactic magnetic field that is organizing the grains (Figure 21).

Within the interstellar gas major structural features have been identified, explored and defined by radio astronomical observations. The most common component is the intercloud gas, found between the spiral arms. It has a kinetic temperature 500–600 K, a density of 2×10^5 atoms per cubic metre and an electron density of 3×10^4 electrons per cubic metre. Embedded in this overall foundation are numerous cloudlets with diameters of roughly 1 pc, masses of up to 10 solar masses, a particle density of 10^7 per cubic metre and a temperature of 30–400 K. These cooler, denser cloudlets may be caused by some general instability in the intercloud gas. In the spiral arms, near to the sun, many dark clouds are present. They have temperatures of 5–20 K, masses of 100–10,000 solar masses and diameters up to 6 pc. Finally, there are the great molecular clouds or black clouds, with temperatures of 30 K, hydrogen densities of 10^9–10^{13} per cubic metre and masses ranging up to one million suns. Within any of these basic structures H II regions will form near to a source of ionization such as an 0 star. In a ring around the galactic centre starting at 4 kpc and going out to 8 kpc H II regions are particularly common.

Our universe is bathed in a weak flux of radiation that has a characteristic temperature of only 2.7 K. The temperatures quoted above are much higher than heating by the background radiation can account for and so it is necessary to consider how the interstellar medium is heated. Galactic starlight does not feed in enough energy to heat all the medium, although in specific cases stars may heat up and ionize substantial volumes of space. Many other forms of heating have been put forward, such as X-ray and cosmic-ray heating, and energy release through cloud collisions.

5.3 Sites of star formation

The subject of star formation received new impetus when molecular clouds and giant H II regions were discovered by radio astronomers. There exists an excellent correlation of the location of H II regions, dust and molecular clouds, with youthful stars such as the T Tauri variables and the O stars. Not until radio and infrared maps of the interstellar medium became available was it realized that such a strong link existed. Stars form from the interstellar gas clouds, but in order for this to happen parts of the gas cloud must fragment into denser clumps known as protostars. This process of fragmentation is almost a total mystery.

The basic physics of how a collapsing protostar turns itself into a new star on the main sequence of stellar evolution was worked out by Hayashi and collaborators. He computed the behaviour of condensing gas spheres, allowing for energy transport by radiation and convection inside the protostars. The initial luminosity can be surprisingly high because in the early stages of evolution the stellar material is transparent to radiation, and so energy can be disposed of rapidly by radiation as the star collapses under its own gravity. Eventually the star becomes opaque and it then traps radiation within itself, so that the luminosity decreases. The trapped radiation destroys molecular hydrogen and ionizes atomic hydrogen (the two principal substances in the protostar). These processes trigger further rapid collapse, followed by an increase in luminosity. Rapid collapse halts only when all the material is ionized and the star enters the final approach to the appropriate point, determined by its mass, on the main sequence.

The sites of star formation in the Galaxy can be determined by looking for young stars. Among these are the O stars and T Tauri stars. O stars have a high luminosity of about 10^4 times the sun, and masses 50 times as great. Energy is consumed so fast that they are on the main sequence for 10 million years or less; this is short compared to the time for one galactic rotation, and so the O stars are still located more or less in the region where they formed. Direct observation and theoretical inference show that these stars are formed in spiral arms and regions that are rich in gas and dust. Also the T Tauri stars, which are believed to be new objects approaching the main sequence, are found in association with dense concentrations of dust. The T Tauri stars are shedding mass and returning matter to the interstellar medium. Probably the mass-loss procedure is an essential step on the route to establishing an equilibrium configuration during the approach to becoming a conventional star. Within the giant H II regions, energy balance calculations demonstrate that luminous O stars are probably the source of energy that ionizes the hydrogen. Compact H II regions may well be sites of star formation that are only 10,000 years old, which is extremely youthful by astronomical standards.

Observations of strong infrared radiation from dust in H II regions and powerful signals from molecular radio sources have a bearing on the star formation problem. As an interstellar cloud condenses and forms fragments, the temperature in the condensations will rise. This is in accordance with the well-known observation that when a gas is compressed its temperature rises. In the astronomical case it is the gravitational force within the gas cloud that causes the compression. However, the rise in temperature is a severe inconvenience to a collapsing protostar because the higher temperature leads to greater internal pressure, and this pressure rise will tend to counterbalance the contraction forces. Somehow the protostar must get rid of thermal energy if stars are to be formed. It happens that prodigious infrared and molecular emitters are seen in just the places where star formation is thought to be occurring. Consequently, it seems reasonable to sketch a picture of star formation, according to which the energy liberated during collapse excites emission by dust and molecules. This disposal of the released energy permits further collapse to take place and the infrared and radio signals to continue. Eventually, stable stars on the main sequence exist, and if any of these are very hot objects they will ionize the interstellar matter and create H II regions.

Throughout its life a star is probably feeding material back into interstellar space, mainly by means of stellar winds. A star's outermost layers are lost continuously, and during the lifetime of an average star its mass is probably decreasing. Within our Galaxy, Edwin Salpeter has estimated that one solar mass per year is being returned from stars back to the interstellar medium. The rate of star formation can be crudely guessed at, and this may be using up one solar mass per year throughout the Galaxy. Although these two estimates indicate that the gas and stars are basically in equilibrium, this is probably wrong, in fact, because both values are only rough estimates. Nevertheless, it is unlikely that there is a very big imbalance between the rates at which new stars use up the interstellar gas and existing stars are replenishing it by means of stellar winds.

In a general way, these remarks on star formation in the Milky Way may be considered to apply to most normal spiral galaxies. The elliptical galaxies, however, tell us a different story. They are virtually devoid of interstellar gas and dust, at least by comparison with the spirals. Their star populations are ancient and well evolved, and there are no new stars being formed in any large numbers at the present time. Here we have a striking astrophysical difference between elliptical and spiral galaxies. Senility characterizes the former, whereas in the latter there seems to be a continuous creation of youthful stars, and an interchange between stellar and interstellar matter.

5.4 The radio emission from spiral galaxies

Viewed from outside, our own Galaxy would appear in the sky as a comparatively weak radio source. Radio observations have led to the discovery of an extended component of radio emission associated with the plane of the Milky Way. Measurements of the spectrum of the galactic background emission reveal that it has a non-thermal origin. Most probably it is due to the synchrotron process; interstellar electrons interacting with the galactic magnetic field create the radio emission. Additionally the gas within the Galaxy has the discrete spectral line radiation from hydrogen at 21 cm, and also the many molecular lines. The disc of the Galaxy also has a certain amount of radio emission that is purely thermal in origin. Spectacular radio sources such as supernovae remnants and pulsars do not contribute significantly to the grand total of radio emission within our Galaxy. At a wavelength of 21 cm the brightness temperature of the Galaxy—a measure of its radio strength—is 0.6 K.

On account of its proximity the Andromeda Nebula (M31) is not difficult to detect with a radio telescope, although it is a weak source. The brightness temperature at 21 cm is only about 0.1 K. The first high-resolution maps of M31 were made by Guy Pooley with the One Mile Telescope at Cambridge, and the most interesting feature is that these show noticeable radio emission from the spiral arms (Plate 6). Also the nucleus of M31 was detected as a feeble radio source. In 1885 a supernova outburst occurred, giving rise to a star which could be observed optically for a time. Supernovae explosions are rare, none having been observed in our Galaxy since the invention of the optical telescope. Radio astronomers might have had an opportunity to investigate the radio waves from the supernova remnant associated with S Andromedae, but it was not detected by Pooley. It is thought that this lack of success may indicate that the gas in the supernova remnant has not yet had time to expand sufficiently to become transparent to radio waves.

Because the galaxy M33 is almost face-on, it was hoped that

especially good radio maps of the spiral structure could be obtained. However, synthesis maps produced in England and the Netherlands have failed to reveal any prominent radio arms, although radiation from the many clouds of ionized hydrogen distributed over the galaxy is detectable.

Spectacular radio maps have been obtained of M51 by the Dutch astronomers. This system is almost face-on and it has a beautiful pattern of optical arms (Plate 7). In fact, it was this galaxy which first revealed the whirlpool pattern of spiral structure as a result of observations by Lord Rosse. The maps made at Westerbork show a series of radio arms that almost coincide with the optical pattern. The main difference is that the radio arms have enhancements of intensity along the inside of the optical arms; this amplification may be due to the compression of interstellar gas in M51 by shock waves generated by the motion of the spiral structure through the galaxy. Another case of a radio spiral is NGC 4258, but this galaxy has radio and optical arms that do not coincide. The reason for this is mysterious, but it could possibly be caused by explosive phenomena.

Two galaxies, NGC 891 and 4631, exhibit faintish radio haloes. For many years controversy has raged over whether or not our own Galaxy is surrounded by an enormous spherical halo, 50,000 kpc across, populated by energetic particles and magnetic field. Certain results in cosmic-ray physics and radio astronomy can be explained more easily if such a halo exists. However, there is no direct experimental evidence of its reality. The Soviet astronomer V. L. Ginzburg made the suggestion that one might look at edge-on spiral galaxies similar to the Milky Way and see if any of these possessed faint radio haloes. Pooley looked at NGC 891 with the Cambridge telescope in 1969, but did not find any spherical source of radio emission outside the plane of NGC 891. Some years later the Dutch astronomers mapped NGC 891 at three wavelengths, 6, 21 and 50 cm, and found that emission at 50 cm could be traced out to 6 kpc above and below the plane, clear evidence for a halo. NGC 4631 is an edge-on Sc galaxy with an extended halo going out to 12 kpc, as well as with emission from its disc.

The nuclei of several spiral galaxies contain radio sources. Our Galaxy possesses a most interesting nuclear component that consists of many giant gas clouds, and mention has already been made of M31. Radio diameters of the nuclei range up to 500 pc with a mean of 200 pc. The fact that some galaxies have radio emission from their discs but do not have a detectable nuclear component could indicate that the nuclei are variable.

There is a weak correlation between the absolute optical magnitude of a spiral galaxy and its radio emission, in the sense that the optically bright objects tend to have higher radio power. Surprisingly, there are no significant differences between the radio powers of isolated and interacting systems. This important result shows that tidal distortion among neighbouring galaxies does not necessarily cause enhancement of the radio emission, even though the interaction might lead to compression of gas and magnetic field. The interacting pair of galaxies NGC 4038/4039 has no radio emission detectable in its long tails, so these cannot consist of threads of hot gas contained by a strong magnetic field as might have been expected for a tidal disruption.

5.5 What kind of Galaxy do we live in?

The morphological type of our own Galaxy cannot be determined with ease because of the simple fact that we live inside it. This makes it difficult to trace out the pattern of the spiral arms for the Galaxy as a whole, especially since obscuration in the galactic plane means that only stars in the nearest arms are observable. There are three properties that favour the view that the Milky Way is an Sb system: the arm shape derived from the radio astronomical observations of hydrogen at 21-cm wavelength; the fraction of neutral hydrogen in the Galaxy; and the fact that wide-angle pictures of the Milky Way reveal a well-developed bulge in the nuclear region. However, there are at least as many observations favouring an Sc classification: optical studies of the spiral arms reveal a more open structure than the radio picture gives; the density of the youngest stars—the O-type —is a pointer to the rate of star formation and in the Milky Way it is nearer to that of M33 (Sc) than of M31 (Sb). A further factor typical of Sc systems is the existence of giant clouds of ionized hydrogen (H II regions). They show up particularly well in radio maps of the galactic centre region, where there may be a dozen or so. Giant H II regions are also found in the disc of the Galaxy and this again is an Sc characteristic. The maximum rotational velocity in the Galaxy exceeds 200 km/s, but in an Sb galaxy the maximum would not be as high as this. If the Galaxy is compared with M31 (Sb) and M33 (Sc) the salient points from a structural standpoint are that the mass is closer to that of M31, and that the number of globular clusters is around 200 for M31 and the Milky Way, but less than 10 for M33. The radio observations of the nuclear zone of the Milky Way show a complex dynamical situation. It is possible that some of the gas clouds there form a barred structure, a factor of importance.

It can be seen that at the present time it is not known with any certainty whether we live in an Sb or Sc galaxy, and the existence of a central bar is not excluded. To a certain extent, this may be merely a semantic issue as not every galaxy can be unambiguously assigned a Hubble type.

5.6 Comparison between our Galaxy and the Magellanic Clouds

Optical studies of the Magellanic Clouds (Plate 19) have been of great importance in the history of modern astronomy. This is because the Clouds are sufficiently close so that stellar astronomers can make detailed studies of the stellar populations and compare the properties of stars in the Clouds with those in the Galaxy. Careful investigation of different star types has shown that the history of star formation in the Clouds has been very different from that in our own Galaxy. This has had a fundamental effect on the relative abundance of heavy elements in the Clouds because the heavy elements are manufactured during the evolution of stars.

Star families are organized according to different rules in the Large and Small Magellanic Clouds. In our Galaxy there are massive globular clusters that probably formed during the collapse phase of the Galaxy, and which are therefore the most ancient stellar organizations present (Figure 22). Also, in the plane of the Galaxy are the younger, loosely structured, open clusters such as h and χ Persei and the Pleiades, which have appeared more recently. Family history has proceeded differently in the Clouds because massive clusters have been forming in a continuous fashion; there are massive, populous clusters such as NGC 121, which is old and highly evolved, as well as mighty, recently formed objects like NGC 1866. Compared to M31 the Clouds have relatively more young clusters, which are luminous and blue on account of their youth; M31 has more ancient clusters. Another oddity about the Cloud clusters is that some of them, and NGC 1221 is an example, are highly flattened. There are no obvious counterparts to these squashed clusters in the Milky

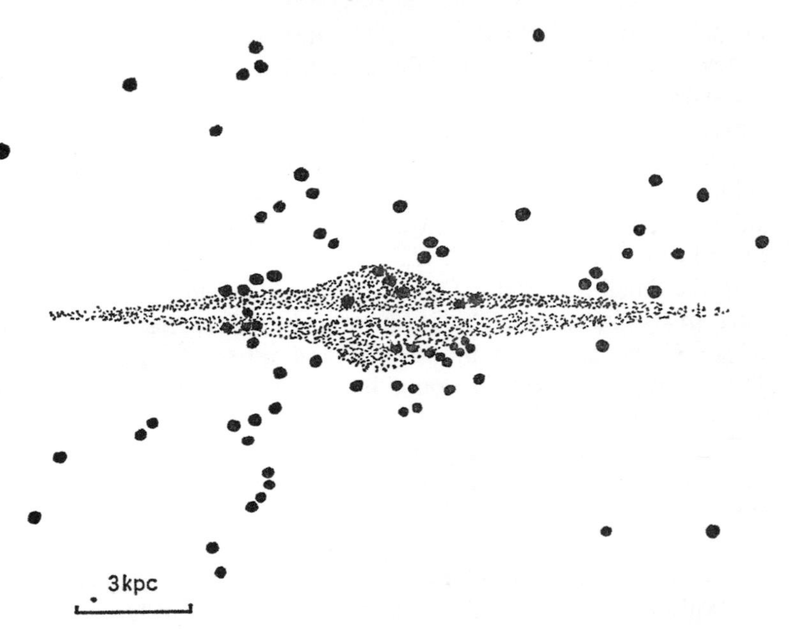

Fig. 22. Schematic representation of the location of globular clusters

Way. Facts such as these are clear evidence that star formation has gone on at different rates in the two systems.

There are also subtle differences in the stars. In the Clouds the red giants tend to be bluer and brighter than red giants in the Galaxy. Our Galaxy possesses giants that are fainter and redder than those found in the Magellanic Clouds. These considerations of the cluster properties suggest that the Small Magellanic Cloud, Large Magellanic Cloud and the Galaxy have all evolved in significantly different ways as regards the stellar component. Even in our local neighbourhood, galactic evolution has proceeded along several paths.

The distribution of stars can also reveal information on the history of a galaxy. Cepheid variable stars are valuable tracers in this respect because the period correlates with luminosity. The latter is determined by the mass of the star, and stars of different mass evolve at different rates. Consequently the spatial locations of Cepheids of various periods can be used to plot the past history of star formation. For Cepheids in the Large Magellanic Cloud it has been found that the younger Cepheids (those with the longest periods of variability) are uniformly distributed. Those of intermediate age show a tendency to concentrate towards the central bar of the Cloud. Finally, the oldest Cepheids, about 50 million years old, are concentrated almost entirely in the central bar. The distribution of Cepheids of different periods shows that the site of recent star formation has shifted drastically from the central bar out to the whole of the dwarf galaxy. It is not understood how such fundamental changes can be caused to occur.

In the Small Magellanic Cloud, there has probably been a recent burst of star formation. This is hinted at by differences in the extent of the Cloud as shown in infrared and ultraviolet photographs. These demonstrate that the young blue stars are mainly congregated in the core of the galaxy.

There are slight systematic differences between the properties of novae in the Magellanic Clouds and in our own Galaxy. These effects are due to small differences in chemical composition, for the Cepheids in the Clouds are relatively deficient in heavy elements. The existence of differences in chemical composition is worrying because the Cepheids act as one of the standard candles for calibrating the distance scale. If the candles are not of a uniform composition throughout, then the distances might be in error. There are behavioural peculiarities associated with the novae that have been observed in the Clouds. In the Galaxy and M31 novae exhibit a tight correlation between the luminosity at maximum light and the decay time. Novae in the Clouds follow the same law if the distance modulus is assumed to be 19.3 mag. However, many methods for determining how far away the Clouds are give a distance modulus of 18.6 ± 0.1 mag. The higher value for the novae is significant and it indicates that nova outbursts in the Clouds are systematically brighter, by about 0.7 mag, than those in the Galaxy and M31.

From a systematic study of the characteristics of stars and star families in the Magellanic Clouds, the overwhelming conclusion which emerges is that the stars in the Clouds contain proportionately fewer of the elements heavier than helium.

6 · Galactic interactions

6.1 Groups of galaxies

The interactions of galaxies with each other are of special interest for several reasons. Mathematically they can present challenging problems: the mutual gravitational attractions of the stars in groups of galaxies that are interacting have to be computed for large numbers of objects. In astrophysics the interacting galaxies have attracted greater attention than their relative numbers might demand because certain groups appear to possess excessively large amounts of kinetic energy.

One obvious type of interaction which we can see is the tidal distortion created by gravitational forces when two large galaxies are separated by only a few times their own diameters. Similarly, unusually powerful activity in a galactic nucleus can fundamentally disturb the outward appearance of a galaxy, or sufficient matter may even be expelled from it to affect nearby galaxies. Objects with strange or abnormal morphology offer the opportunity to view the dynamical effects of nuclear instabilities, immense gravitational forces, and titanic explosions on the external appearance of the galaxies. Physical conditions that could never be reproduced on Earth can be studied in the interacting and peculiar galaxies.

Modern work on contorted galaxies commenced with the availability of the Palomar Observatory Sky Survey, which made it possible to catalogue peculiar galaxies by scanning the newly available prints for oddities. In 1959 B. A. Vorontsov-Velyaminov published his *Atlas of Interacting Galaxies*, which lists many examples of close groups of galaxies that seem to be interfering with each other. Individual cases cited from this catalogue are prefixed by the code VV followed by the catalogue number. Thus, VV 172 refers to the interacting group listed under entry 172 in Vorontsov-Velyaminov's catalogue. Halton Arp has compiled an *Atlas of Peculiar Galaxies*, published in 1966, and Fritz Zwicky produced a compilation of eruptive and post-eruptive galaxies. These latter are suspected of suffering gross explosions.

Interaction and peculiarity manifests itself in many different forms in these catalogues of strange cosmic beasts. Among the dynamically interesting groups are chains and lines of galaxies suspended in the sky like strings of beads. In the Perseus cluster of galaxies there is the Markarian chain, a series of two dozen great starry systems strung across 500 kpc or more of space (Plate 8). Other linear associations tions consist of handfuls of objects almost in contact; examples are VV 172 and Arp 330 (both discussed below). Apart from the chains there are also tight clusterings, such as Stephan's quintet and VV 282, in which the different velocities of individual members indicate that the compact associations will probably dissolve on a time scale of a few hundred million years. Another category of devious behaviour takes in the galaxies which appear to be tearing each other apart, by rending each other's structure while they are locked in gravitational conflict.

6.2 Runaway galaxies break the chains

Before any discussion of instabilities in groups and clusters of galaxies can get under way we need a criterion which defines the stability of the system. Galactic interactions tend to be a battle, generally an uneven one, between two types of energy. On the one hand the kinetic energy of a given galaxy will tend to move it away from its companions by virtue of the relative motions involved. Kinetic energy, therefore, distributed among group members, tends to break up the group over time. However, we must also take gravitational forces into account because these will try to pull galaxies closer and closer together. This dynamical tug-of-war is quantified through a powerful statement known as the virial theorem. For our purposes we can phrase this theorem thus:

> If a group of gravitationally interacting bodies is in a stationary state, then the sum of *twice* the kinetic energy and the gravitational potential energy is zero.

Symbolically we can write this as

$$2T + V = 0, \tag{1}$$

where T represents the kinetic energy bound up in the motion. V is the potential energy stored in the gravitational field and it is negative. If the sum in (1) is less than zero (V large and negative) for a particular group of galaxies, gravity is in full control and the group will remain closely knit. However, should the sum in (1) be positive then the velocities of individual members are too great for gravitation to restrain the tendency to disperse, so the group will ultimately break up.

If we want to use the virial theorem as a stability check on a particular group it is first necessary to estimate T and V. Both of these quantities involve the masses of the galaxies, which are difficult to find. Consequently the argument is usually inverted and we ask what mass a group must have if it is to be stable. The problem then is to observe the velocities of group members, and their positions relative to each other. Once these data are available it is possible to calculate, via the virial theorem, the mass required to keep everything bound, that is to ensure that the sum in (1) above shall not be a positive quantity. If the binding mass so derived far exceeds the sum of the likely masses of individual members, then the configuration under investigation is, in some sense, unusual and demands a special explanation.

Wallace Sargent has examined the velocities of the members of groups and small clusters of galaxies and he found cases where the masses deduced by application of the virial theorem are much larger than those masses actually observed as luminous matter. Group VV 159, for example, has three members. Spectra show that two of these have radial velocities of 10,500 km/sec with respect to our Galaxy, whereas the third member is tearing away at 13,200 km/sec. The VV 159 grouping is obviously dispersing rapidly. The two slower galaxies do not pull enough weight to keep the faster member, speeding away from their grasp at 2700 km/sec, within reach for much longer. Note, however, that the maverick may not necessarily be related to the two 10,500 km/sec galaxies. It could be a more remote galaxy, at a distance where the velocity of recession is 13,200 km/sec, that happens by chance to appear in the same field of view as the two 10,500 km/sec objects.

A more intriguing group is known as Arp 330 (Plate 10). This well-defined system of half a dozen objects encompasses a velocity spread of 1200 km/sec, which is too large to ensure long-term stability. In fact, the system will break up and the members go their separate ways over the next 200 million years.

A puzzling runaway galaxy is found in the chain VV 172 (Plate 9). In 1968 Wallace Sargent made the astonishing discovery that one member has a recession velocity of 36,880 km/sec, whereas its companions are coasting along at 16,000 km/sec. How are we to account for a difference of nearly 21,000 km/sec in the velocities of group members? A cautious investigator would first consider the possibilities of a freak coincidence or projection effects. Perhaps the odd man out is a far-off, high-luminosity galaxy that is seen purely by chance in projection against the chain of four nearer galaxies. Sargent has set odds of 5000 to 1 on such an arrangement being a caprice of nature.

In 1951 Carl Seyfert pondered a clump of extragalactic objects now called Seyfert's sextet (Plate 12). It must be the most unharmonious sextet in the universe. Five of the members are spirals and the sixth is an irregular cloud that may have become detached from one of the spirals. Brightnesses of these six vary from 14.7 to 16.5 mag, a range that does not permit a huge spread of distance. One of the sextet members has a severely discrepant velocity because it is shooting away at 19,930 km/sec, in conflict with the mean velocity of 4400 km/sec set by the other members. The mean separation of the members is a mere 17 kpc, a value which indicates the close association of the six galaxies. Seyfert himself stated that the probability that the sextet is an artefact created by a chance overlap of objects at greatly different distances—and therefore characterized by different redshifts—seemed extremely remote.

Halton Arp has worked indefatigably on the problems of interacting galaxies. In 1973 he claimed that the observed sizes of the H II regions and bright emission clumps of the discrepant velocity object in Seyfert's sextet precluded its being a remote background object. According to Arp the fact that its internal structures can be clearly discerned demonstrates that the 19,930-km/sec galaxy must be closer to us than its redshift would normally indicate. This argues strongly that the high-velocity object is a member of the sextet. The tale of Seyfert's sextet has grown curiouser and curiouser over the years: 1.5° from the six is a large galaxy. This is extremely peculiar, and its morphology gives the impression that it has suffered some tremendous explosion. Its velocity of recession is 4600 km/sec, almost an exact match for the velocities of the majority of the population of Seyfert's sextet. Probably this is only a coincidence, but we should keep our minds open to the possibility that the sextet and the tortured giant are physically related.

The most celebrated group of interacting galaxies is known as Stephan's quintet (Plate 11). The history of this group goes back to 1877 when M. E. Stephan, using the Marseilles telescope, observed four galaxies that were close together in the sky. Later, photographs showed the four to be very distorted and one of them to be double. In 1956 redshifts for four of the members became available and these ranged from 5700 to 6700 km/sec. On the basis of these values Ambartsumian argued that the system must be dynamically unstable, and this caused further interest. Then in 1961 Margaret and Geoffrey Burbidge came up with a real puzzle; they had observed NGC 7320, a low-brightness member of the group and they found that it had a velocity of only 800 km/sec. Soon afterwards, at a conference in Santa Barbara, California, on the stability of systems, they put forward two hypotheses to account for the difference in redshift of nearly 6000 km/sec. Either NGC 7320 is an unrelated foreground galaxy, or NGC 7320 has been expelled explosively from the group. The chance of coincidence was put at 1 in 1500 and the explosive hypothesis clearly lay outside the bounds of conventional galactic dynamics.

Ron Allen brought radio astronomical observations to bear on the problem of the quintet in 1970, with a study of the distribution of neutral hydrogen in NGC 7320. He found that NGC 7320 should be fairly close in order to avoid an implausibly large hydrogen mass for NGC 7320. The issue at stake here is whether or not NGC 7320 is at the same distance as the high-redshift members. This question assumed greater importance when the intriguing problems associated with the groups mentioned above came to light. Arp went so far in 1970 as to argue that all parts of Stephan's quintet have been blasted out of an additional object in the same field, the large spiral galaxy NGC 7331 (redshift 800 km/sec). He found, in 1972, evidence from radio emission in the vicinity of the quintet to back his argument.

In 1973 Arp published observations of the quintet made over a number of years with the 5-metre telescope of the Hale Observatories. These observations were aimed at finding the distances to the quintet of galaxies without resort to statistical or redshift-based arguments. From a study of the diameters and distribution of H II regions he argued that two of the high-velocity members are at the same distance as NGC 7320; furthermore, he claimed that the two high-velocity objects were not as far away as the cosmological distances corresponding to velocities of 6,700 km/sec. Strangely enough, however, he claimed that one of the 6,700 km/sec galaxies probably is at its cosmological distance. Clearly there is something very strange afoot in the quintet.

The contradictory picture becomes even more confused when the overall morphology of the group is considered. A filament of bright hydrogen emission threads the quintet. And radio mapping with the synthesis telescope at Westerbork in the Netherlands adds further evidence for interaction among the galaxies. Arp concluded, on balance, that all the members of the quintet are interacting and are at a close distance of 10 Mpc.

Jurgen Materne and Gustav Tammann analysed the dynamical properties in 1974. They came to the conclusion that the four high-velocity members constitute an independent stable grouping, and that NGC 7320 is a nearer galaxy in the foreground which is not related to the other four.

Stephan's quintet, like Seyfert's sextet, is close to a giant spiral, NGC 7331. In a study of many groups of interacting galaxies, published in 1973, Arp indicated that the multiple-interacting groups fall close to giant galaxies. This is taken by him to indicate that the interacting groups have their origin in larger galaxies. This is a revolutionary view, not shared by most astronomers. Convincing though the circumstantial evidence appears to be, there is a general feeling that it could all be a matter of coincidences, projections and chance overlaps; by 1975 several researchers were inclined to think that the number of unstable groups had been over-estimated. It could be argued that we do not have sufficient data on the structure and dynamics of small groups to decide the issue unequivocally. If, however, the contorted groups do originate inside other galaxies, then the generally accepted picture of the formation and evolution of galaxies would need drastic revision. It is therefore of considerable importance to find out more about the interacting galaxies in order to decide whether or not there is something physically mysterious happening.

6.3 Bridges and tails between the galaxies

The galaxy catalogues of Arp, Vorontsov-Velyaminov and Zwicky contain many examples where tidal forces between galaxies may be responsible for at least some of the peculiarities. Often ribbons and filaments of matter are observed streaming out of one galaxy, towards another, across vast reaches of intergalactic space. Sometimes these outstretched luminous fingers link up and bridge the voids. Occasionally the arms of neighbouring spirals are seemingly entangled. When two mighty galaxies seem to be colliding, a rare event, great streamers are seen thrust out, from the confused central mass, to span many tens of kiloparsecs. The streamers occur under two circumstances: there are various bridge arrangements where a spiral arm in a large galaxy links into a subordinate companion, and there are also faint but extensive tails that are attached to certain of the multiple galaxies. The great classical example of a spiral arm bridge is that linking M51 to NGC 5195 (Plate 7).

The physical origin of the bridges and tails has been hotly disputed. Intuitively, we anticipate that mutual gravitational attractions may be capable of drawing threads of stars and gas out of neighbouring galaxies. Many theorists, however, prefer to ascribe the bewildering variety of features to explosive and cataclysmic origins. With the advent of more powerful computers, however, it became possible to put the matter to quantitative test.

This is what Alar and Juri Toomre have done in their many theoretical investigations of the tidal interactions between galaxies. For computational simplicity they have considered the encounters of pairs of disc-like galaxies; each galaxy is idealized as a disc of particles, and the forces between the particles in one galaxy and the other evaluated. Although these models may appear grossly simplistic, the interactions between them probably mimic the real world reasonably well, and many types of close passage between model galaxies have been computed. The work has shown that dense and narrow bridges of matter arise tidally if a small satellite galaxy passes close to a massive primary galaxy. Distortions provoked by gravity then pull the outer portion of the disc of the primary into a filament linking both objects. On the far side of the larger galaxy a counterarm sometimes sprouts out. Long tails are pulled out when two objects of nearly equal mass pass each other at a close distance and slowly at that. Then a great streamer of escaping debris is pulled out. In this way extremely thin tails might be explained, because they could be normal ribbons of debris, produced tidally, but viewed from the side.

The Toomres have claimed that the spectacular bridge apparently uniting M51 to NGC 5195 is a great cosmic hoax. Both of these galaxies show extensive signs of tidal damage; in particular, the arms of M51 are deformed at the outer parts of the galactic disc. An important clue in piecing together the past history of this interacting pair is that NGC 5195 has a velocity which is 110 km/sec greater than M51. This can be interpreted as evidence that NGC 5195 is a satellite galaxy in orbit around M51. The luminous link between the two can be explained as an artefact of a close passage that took place 50 million years ago. This has drawn out a streamer of matter; NGC 5195 is now many tens of kiloparsecs beyond M51, but appears behind the tidal filament. Seen in projection it is thus apparently joined to M51, but in reality is quite separate. The time for a full orbit of the satellite M51 is set at 2.5 billion years.

In 1973 three Australian radio astronomers, D. S. Mathewson, M. N. Cleary and J. D. Murray made observations that led to the discovery of a spectacular tidal interaction on our own doorstep. From measurements of hydrogen at 21 cm they found a delicate filament of hydrogen gas extending from the Magellanic Clouds down to the south galactic pole, on one side, and right across the galactic plane on the other (Figure 23). This thread of gas has been spun in a perfect arc across 180° of the heavens. Its mass may total 10^9 solar masses, which is about equal to the combined total of neutral hydrogen in both Magellanic Clouds.

How are we to explain this stream of gas? One possibility is that it is debris left behind the Clouds during their orbit around our

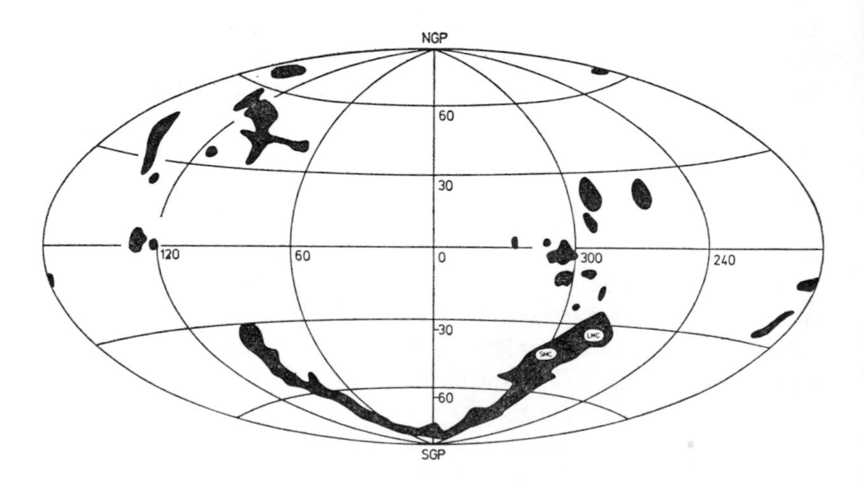

Fig. 23. The Magellanic Stream in galactic coordinates

Galaxy, rather in the manner that comets leave a trail of meteoritic debris in their wake. However, the stream is too well organized to be explained in this way. There are difficulties with keeping a filament of gas stable over sufficiently long time scales.

An alternative interpretation is the genesis of the stream in a mighty gravitational conflict involving the Clouds and our Galaxy. The Small Cloud could have grazed the Galaxy, in a passage only 20 kpc from the nucleus, about 5×10^8 years ago. This would have raised enormous tidal forces and caused the evisceration of hydrogen entrails from the Milky Way. This theory would imply that the Clouds and our Galaxy have, in the past, been parties to the type of interaction which we see taking place right now in more distant objects. One objection to this dynamical model for the origin of the stream is that it is rather unlikely that the Small Cloud has ever, in fact, had such a near collision with the Galaxy.

A third possibility is that the stream is foetal gas remaining after the formation of the Clouds, and that the two galaxies are still embracing it. The stream is then explained as ejecta from the vicinity of the Clouds, the relic being thrown out perhaps by collisions with the intergalactic wind, or else by explosions. The discovery of the Magellanic Stream shows that we should be prepared for exotic interactions even on our own doorstep.

6.4 Exploding galaxies

The beautiful photographs of spirals such as M31 and M51 evoke a universe of serenity and changelessness, the great nebulae apparently wheeling in the peaceful voids. This romantic image was violently shattered with the discovery that, among the great swarms of galaxies there is a significant number in great disarray. We have already indicated the power of tides on a galactic scale and now we consider galaxies which can be utterly wrecked by mighty explosions from within. A classic example is M82 with its fans of gas and dust forming enormous plumes on each side (Plate 13); another is M87, the giant elliptical in the Virgo cluster, whose nucleus has sprouted a great jet of highly energetic matter. Early evidence that galaxies may explode came from radio astronomical studies in the 1950s, and we shall return to this aspect in the next chapter. For the present, the optical evidence for the occurrence of violent events is considered.

Photographs of M82 taken in 1949 with the newly completed 5-metre Hale telescope showed filamentary structures emerging from the central parts of the galaxy. In 1962 Allan Sandage obtained further photographs, this time taken through interference filters designed to let only the red light from hydrogen onto the photographic plate, and so accentuate the gas structures. These pictures of the hydrogen wisps were apparently great fans of gas reaching out 4 kpc above and below the plane of the galaxy (Plate 13). An independent spectroscopic investigation by Roger Lynds revealed that the gas in these plumes was in rapid motion. Lynds' spectra showed not only that the gas was travelling away from the galactic nucleus at speeds approaching 1000 km/sec, but also that the velocity increased with increasing distance from the centre. In this object the velocity of the gas is related to distance in such a way that all the matter must have emerged from the nucleus at one specific time in the past. We are currently viewing a situation in which the matter ejected at the highest speeds has travelled the greatest distances.

According to the variation of velocity with distance, M82 must have erupted 1.5 million years ago. By considering the strength of spectral-line radiation from the filaments it is possible to estimate their total mass; it is about 5 million solar masses, or some 0.05 per cent of the total mass of M82. Considerable energy was required to set this mass on its great journey: about 2×10^{48} joules.

The light from the filaments was shown to be strongly polarized (25–30 per cent) by Visvanathan and Sandage in 1972. At first it seemed possible that this polarization was due to the scattering of light from the brilliant central regions of M82 by dust grains in the plumes. In 1973, however, it was suggested that this is not so and an alternative idea, that the hydrogen gas shines by fluorescence, was proposed. This second hypothesis does not conflict with the explosive origin of the plumes, as does the dust theory.

Additional evidence on excitation in M82 was provided by Sidney van den Bergh. In a study of the central region of the galaxy, he found some half-dozen extremely large and brilliant H II regions. These could be remnants of the great explosion, or could even have helped to stoke the flames in the first place. The nucleus of M82 is a strong infrared emitter, with a total luminosity of 2×10^{37} watts. Observations with high-resolution radio telescopes at the University of Cambridge by Philip Hargrave have revealed a compact radio source close to the optical nucleus (Plate 13). Hargrave believes that a compact object at the nucleus feeds high-speed electrons into the M82 radio source. Furthermore he has advanced the interesting hypothesis that the fans of gas and dust may have been expelled by the radiation pressure exerted by the strong infrared flux.

Another galaxy of great interest in tracing the history of violent events is M87. This object has sprouted a jet which is unique in its observed properties, its nearness to our Galaxy and in the rich variety of observations made of it. The optical jet extends from the nucleus out to a distance of about 1.2 kpc. Short-exposure photographs by Jim Felton, Halton Arp and Roger Lynds have resolved the jet into six condensations, which emit a blue continuum of light with detectable spectral lines. Most of the brilliance comes from remarkably small volumes, for parts of the jet are a mere 70 pc in diameter, which is minute compared to the size of a giant elliptical galaxy. At radio wavelengths M87 is the third brightest source in the sky, and the bright components of the jet make a significant contribution to the total intensity. The light and radio emissions are linearly polarized; this observation suggests that the radiation is caused by the interaction of very-high-energy electrons with the magnetic field in the jet, to produce what is known as synchrotron radiation (Plate 14).

For theorists the existence of the jet presents certain problems. A plasma knot composed of high-energy electrons and magnetic field will tend to expand at a fair fraction of the speed of light unless restrained by external pressure. The shape of the jet suggests that it consists of material expelled from the nucleus, and according to its present length this ejection must have been going on for several thousand years. Yet the thickness is only tens of light years, so there must be a powerful constraint to the natural expansion of the gas. Early bizarre explanations of this containment included the idea that the jet might consist of a series of small, dense objects with masses exceeding a million solar masses. Another possibility is that the condensations contain several million solar masses of cold gas; the torpid influence of this cold gas could restrain the efforts by the high-energy electrons to expand the clouds. Yet a further suggestion is that the M87 jet is a supersonic stream of hot gas which is being produced continuously in the nucleus. Whether these or other models are adopted, they all have a common basis: that M87 possesses a massive plume of hot gas which has been ejected from the nucleus.

6.5 The energy expenders

A constant theme to emerge from studies of a wide variety of chaotic galaxies is the existence of systems that contain vast reserves of apparently surplus energy. We are not certain how the positive energy systems fit into the general picture of galactic evolution. The interacting galaxies have helped to demolish the classical view that the extragalactic world is the embodiment of peace and serenity. Although we can now give a qualitative picture of certain types of interaction, each time we had to pull a rabbit out of the hat—a mysterious source of energy. The evidence that energy is there in abundance is convincing. But we have only begun to scratch the surface in the battle to explain whence this energy has come, and we return to a consideration of this difficulty again in Chapter 12.

On the observational side there is need to extract more information from the galactic odd men out. Spectroscopy of the weird objects in Arp's catalogue would help, because this can provide, ideally, information on velocities, physical conditions and lifetimes. Certain of the conclusions to emerge from the tidal force calculations can also be checked from velocity measurements. Also, there is the vexed question of the validity of statistical arguments, which, it will be remembered, figured prominently in the discussion on runaway groups. Perhaps closer attention should be paid to groups that do not call attention to themselves by peculiarity; are all the objects in 'normal' groupings at the same distances? Any method of finding distances to the members of interacting groups would be of great value.

7 · The nuclei of galaxies

7.1 The central regions of galaxies

In most galaxies the matter density reaches a maximum somewhere near the centre of the galaxy. Many of the nearer objects, such as M31, M33 and M51 have a starlike or almost starlike point of light superimposed on this region of maximum density. At greater distances the limitations on the angular resolution of telescopes do not allow us to discern a starlike centre, because such an image is swamped by background light from the bright central regions. However, in the Seyfert galaxies it is possible to see a brilliant jewel of nuclear light even at large distances. The concentrations of light embedded in the central regions of the galaxies are termed galactic nuclei. They have shown us a number of exceptionally interesting astrophysical phenomena, and the study of them has led to several important advances in modern astronomy.

Spectroscopic analysis of the radiation from galactic nuclei shows that processes take place within the nucleus that are generally absent from the spiral arm regions or the outer parts of elliptical galaxies. For example, according to Doppler shifts in the observed spectra, there are violent motions of gas clouds which are frequently observed travelling at speeds running into hundreds of kilometres per second. Also, the luminosity of some galactic nuclei shows relatively sharp and short-lived variations. Furthermore, the nuclei of many galaxies contain a radio source. Often they show an excess of emission in the ultraviolet, and some nuclei are prodigious emitters in the infrared.

These phenomena can be taken as signs of activity in the galactic nucleus. Frequently this activity is a major source of energy and luminosity in an active galaxy, and the exotic processes that are taking place have received much attention from optical and theoretical astronomers. In fact, all galaxies probably have at least a small amount of unusual activity in their central regions; even the serene Andromeda galaxy, long regarded as having a dead central region containing only low mass stars, is expelling matter at the rate of 0.01 solar masses per year. At the other end of the activity spectrum are Seyfert galaxies and quasars, where the energy output from compact regions totally dominates any emission from ordinary stars.

7.2 The heart of the Milky Way

Despite the fact that our Galaxy's nucleus is the nearest, we are at a considerable disadvantage when we observe it optically, simply because an absorbing layer in the galactic plane diminishes its light intensity by at least 25 magnitudes on its long journey from the galactic centre to the solar system. Consequently, it is practically impossible to make any worthwhile observations of this interesting region with optical telescopes. This huge disadvantage is partially compensated for, fortunately, by the fact that radio, infrared and ultraviolet frequencies can penetrate the galactic haze. Detailed observations of the nucleus are made by radio astronomers, for example, to an extent which cannot be realized for the nuclei of even the nearest galaxies beyond our own.

What, then, do we find at the hub of the Milky Way? Studies have shown that the nucleus is mainly made of stars, with gas and dust contributing only a few per cent of the mass. Information on the amount of matter at the centre, its distribution and its motion comes mainly from studies made by radio and infrared astronomers. Ever since the detection of the 21-cm line of hydrogen, the motion of the gas in the galactic centre, revealed by small frequency shifts that can be attributed to the Doppler effect, has been a topic of prime importance. At Dwingeloo, in the Netherlands, Jan Oort and his associates showed in the late 1950s and early 1960s that large-scale gas motions exist, and that these can be interpreted in terms of the expansion of hydrogen clouds away from the nucleus of the galaxy.

The density of matter in the nuclear regions can be deduced from the velocities of rotation of material in the nuclear disc, on the assumption that the nuclear region of our Galaxy is basically similar to that of nearby M31. With the advent of infrared detectors, which are able to penetrate the great dust clouds that block the visible light, it became possible to estimate the star density right up to the centre of the nuclear disc. There are good reasons for believing that the intensity of radiation at 2.2-μm (2.2–micron) wavelength is proportional to the star density; in M31 the distribution of 2.2-μm infrared intensity and optical intensity are the same. Since the latter comes from stars it is evident that the 2.2-μm radiation correlates with the density distribution of stars. If this correlation holds up for our Galaxy also, then, according to Oort, the radio and infrared data may be interpreted in the following manner: at 100 pc from the centre the density of matter is 105 solar masses per cubic parsec ($M_{\odot}$ pc^{-3}); this is roughly fifty times the density of stars in the solar neighbourhood. As we probe closer to the centre of the nucleus the star density rises rapidly with decreasing radius, following a power law. At 10 pc the density is 6.6×10^3 $M_{\odot}$ pc^{-3}, at 1 pc it is 4.2×10^5 $M_{\odot}$ pc^{-3}, and at 0.1 pc from the centre it is estimated as 2.6×10^7 $M_{\odot}$ pc^{-3}. This last figure is around 100 million times the density of

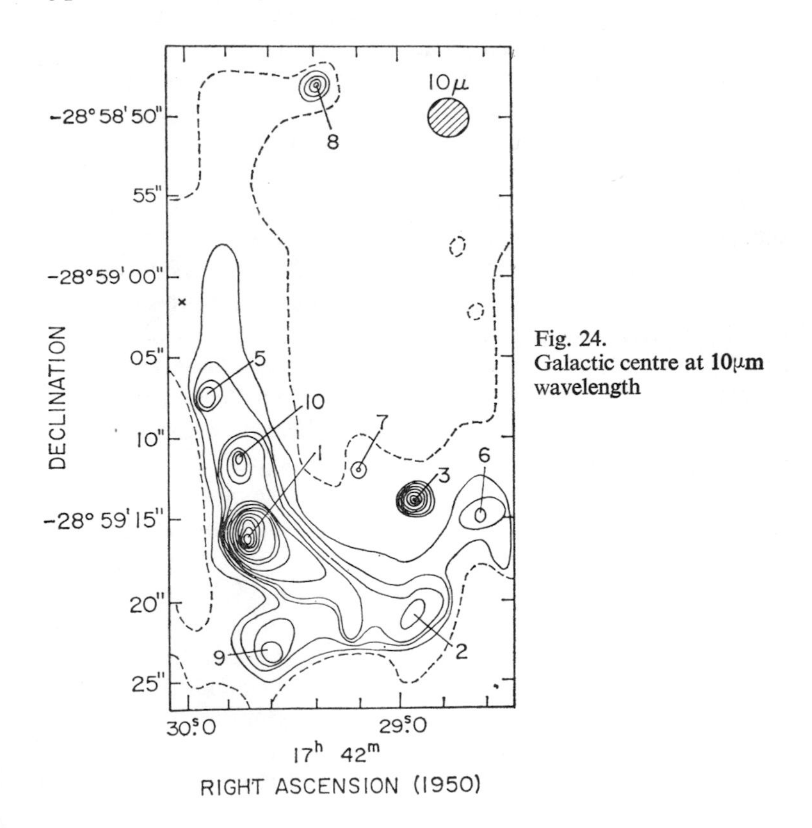

Fig. 24.
Galactic centre at 10μm wavelength

stars near our sun! The total mass enclosed by a central sphere of radius 1 pc is 4.4 million solar masses, and there are 70 million solar masses within 10 pc of the centre. Altogether the mass in the galactic nucleus is roughly 1000 times higher than in a large star cluster, such as M3.

The nuclear disc of hydrogen gas perceived by radio astronomers extends out to 750 pc, has a thickness ranging from 100 to 300 pc, and its outer periphery whirls round at 230 km/sec. This disc also shows up as a very intense infrared emitter at 100 μm, indicating that it contains hot radiating dust as well as hydrogen gas. The amount of gas and dust swirling around in the disc is still largely unknown although contemporary estimates range from 5×10^6 up to 2×10^8 solar masses. Much fine structure is visible in the map of the galactic centre made at 10 μm, as though several discrete regions are each contributing to the emission (Figure 24). Although one cannot yet be certain, it appears likely that the long-wavelength infrared emission from the nuclear disc is generated by dust which has been heated up through the absorption of higher energy optical and ultraviolet radiation. Eric Becklin and Gerry Neugebauer have also reported the detection, at 2.2 μm, of a point source within 10 arc seconds of the galactic centre. If this bright object is indeed close to the galactic centre, then it has a total luminosity at least 10^5 times that of the sun, together with a temperature of about 2000 K.

Considerable circumstantial evidence has now accumulated in favour of the view that explosions have taken place recently in the galactic centre, resulting in the ejection of substantial quantities of gas. The most striking structural feature is an arm of gas some 3 kpc from the nucleus, which is moving away from the centre at about 50 km/sec. On the far side of the nucleus there may be another tube of material racing away at 135 km/sec. Pieter van der Kruit of the Leiden Observatory has discovered considerable masses of gas above and below the plane of the nuclear disc. This material appears to be travelling away from the plane in a manner consistent with cataclysmic expulsion, since the velocities exceed 100 km/sec in certain cases. Altogether at least 10^6 solar masses of hydrogen are participating in the large-scale motion. It is possible to trace back, theoretically, the past history of the moving gas, and thus infer that it could have originated in a stupendous explosion 12–13 million years ago, with the subsequent expulsion of material at initial speeds of 600–700 km/sec. This activity lasted for a few million years, and

1 The 2.25-metre telescope of the Institute for Astronomy, University of Hawaii

2 This 100-metre instrument at Effelsberg, in West Germany, is the world's largest fully-steerable radio telescope

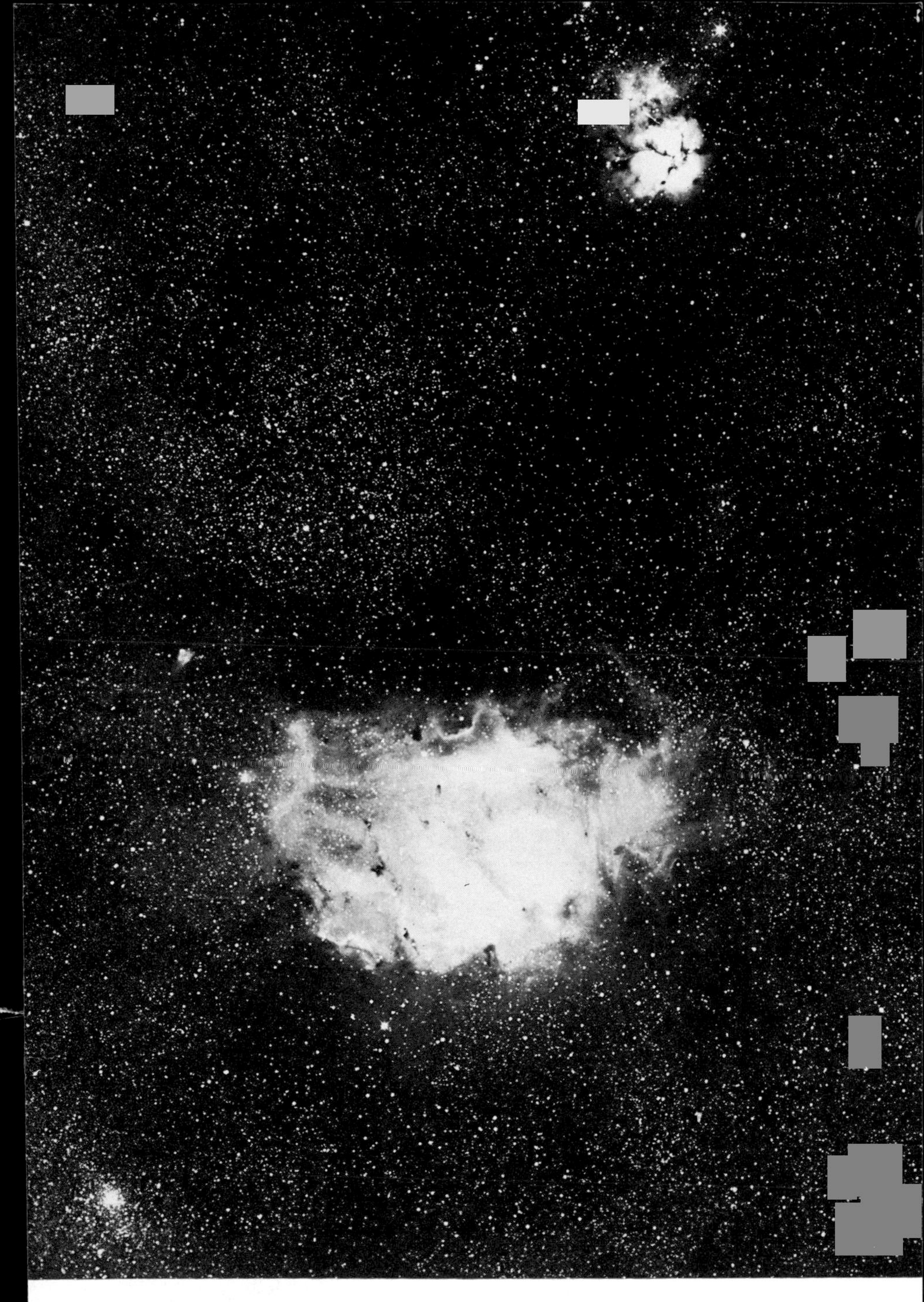

5 The Trifid Nebula is composed of hot gas and obscuring dust

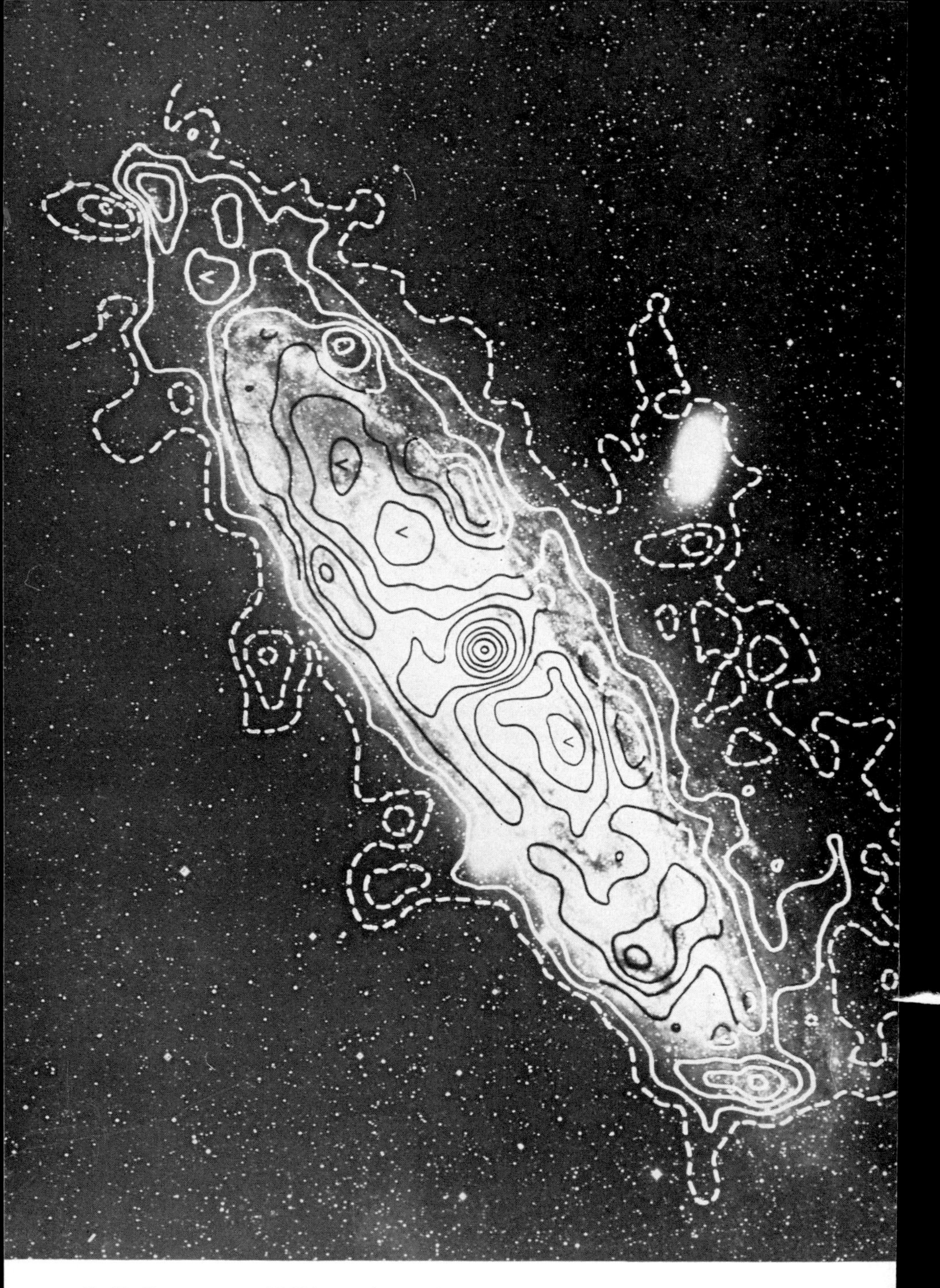

6 Radio contours of M31 superimposed on the optical structure

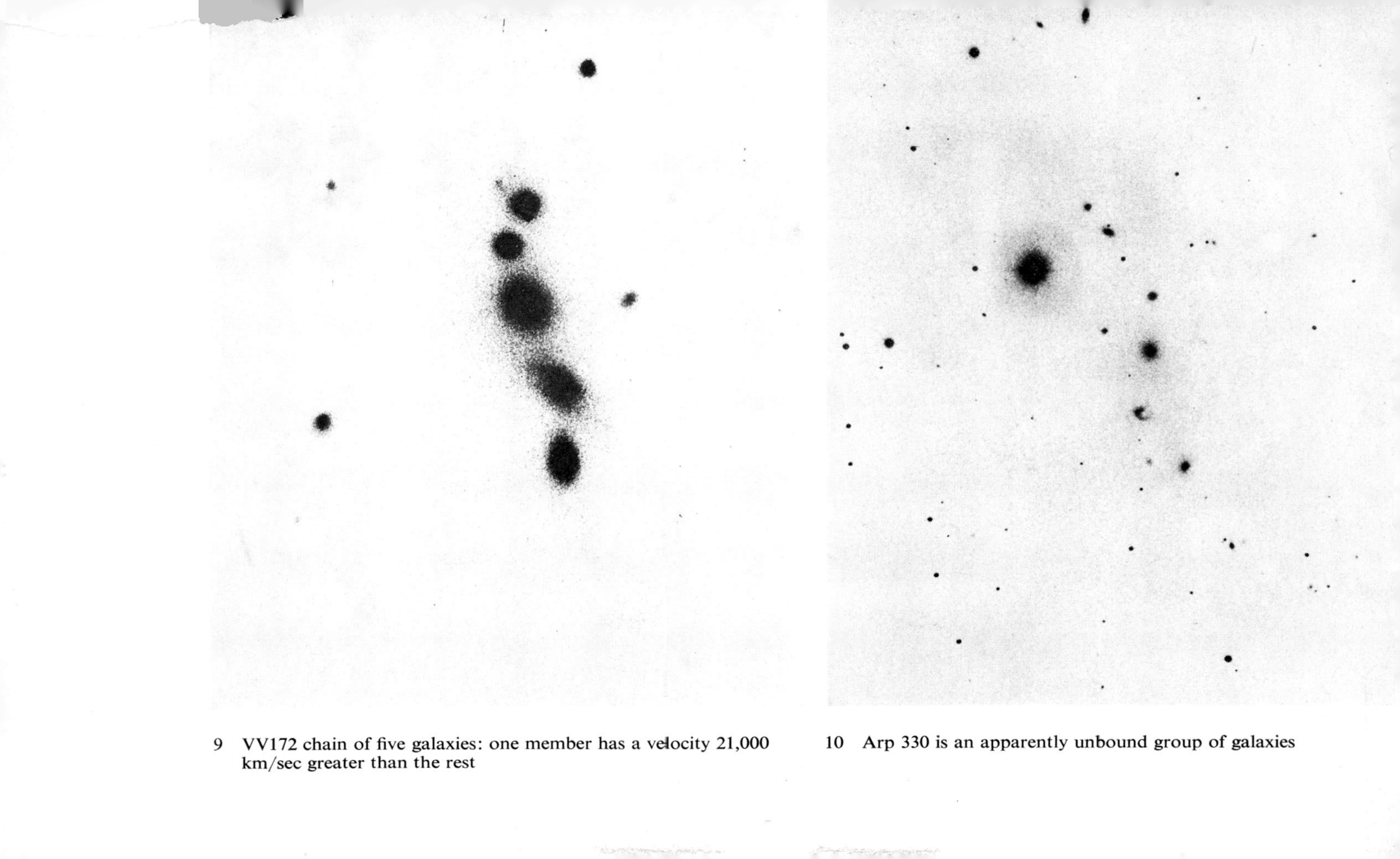

9 VV172 chain of five galaxies: one member has a velocity 21,000 km/sec greater than the rest

10 Arp 330 is an apparently unbound group of galaxies

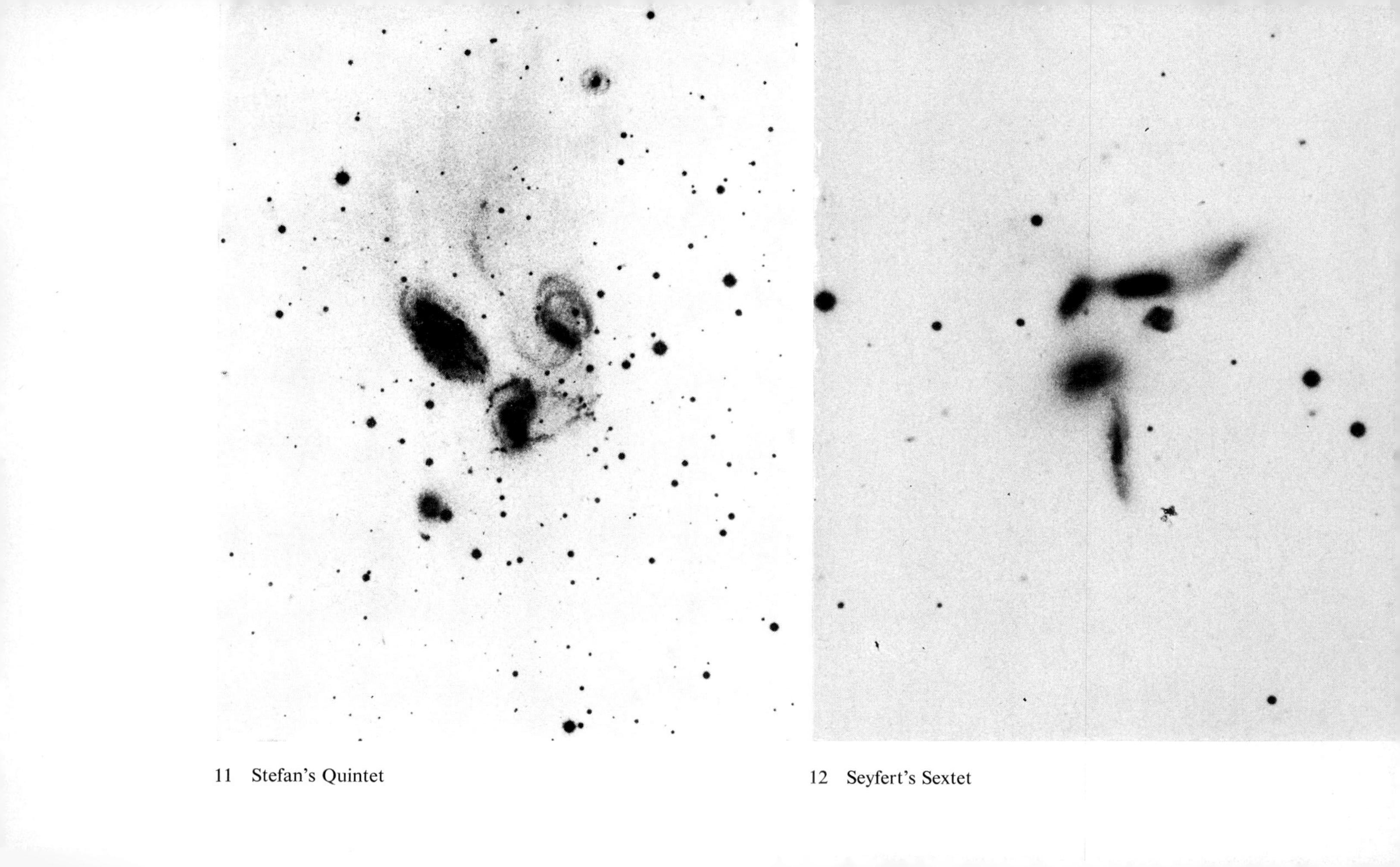

11 Stefan's Quintet

12 Seyfert's Sextet

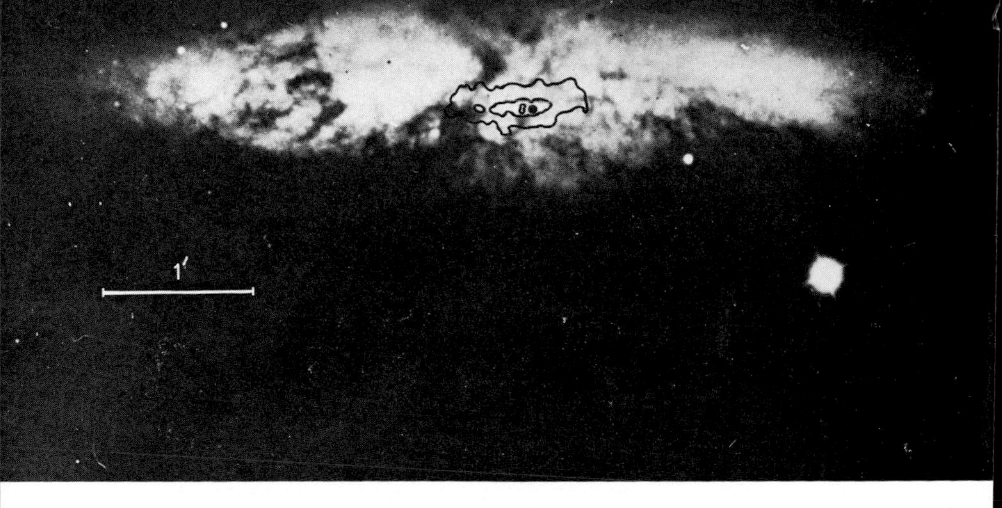

13 Exploding galaxy M82 and its nuclear radio source

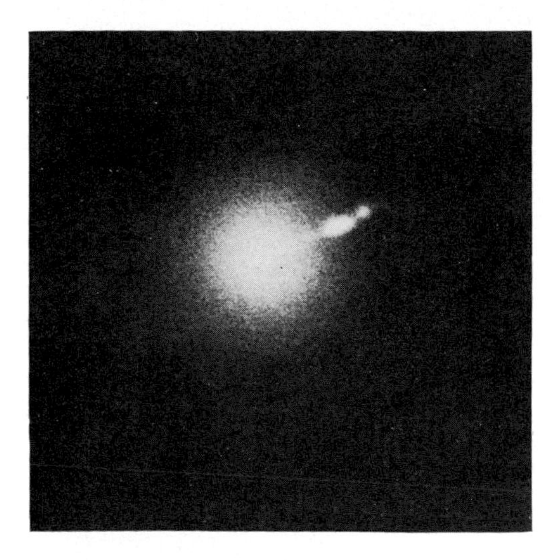

14 The jet of M87

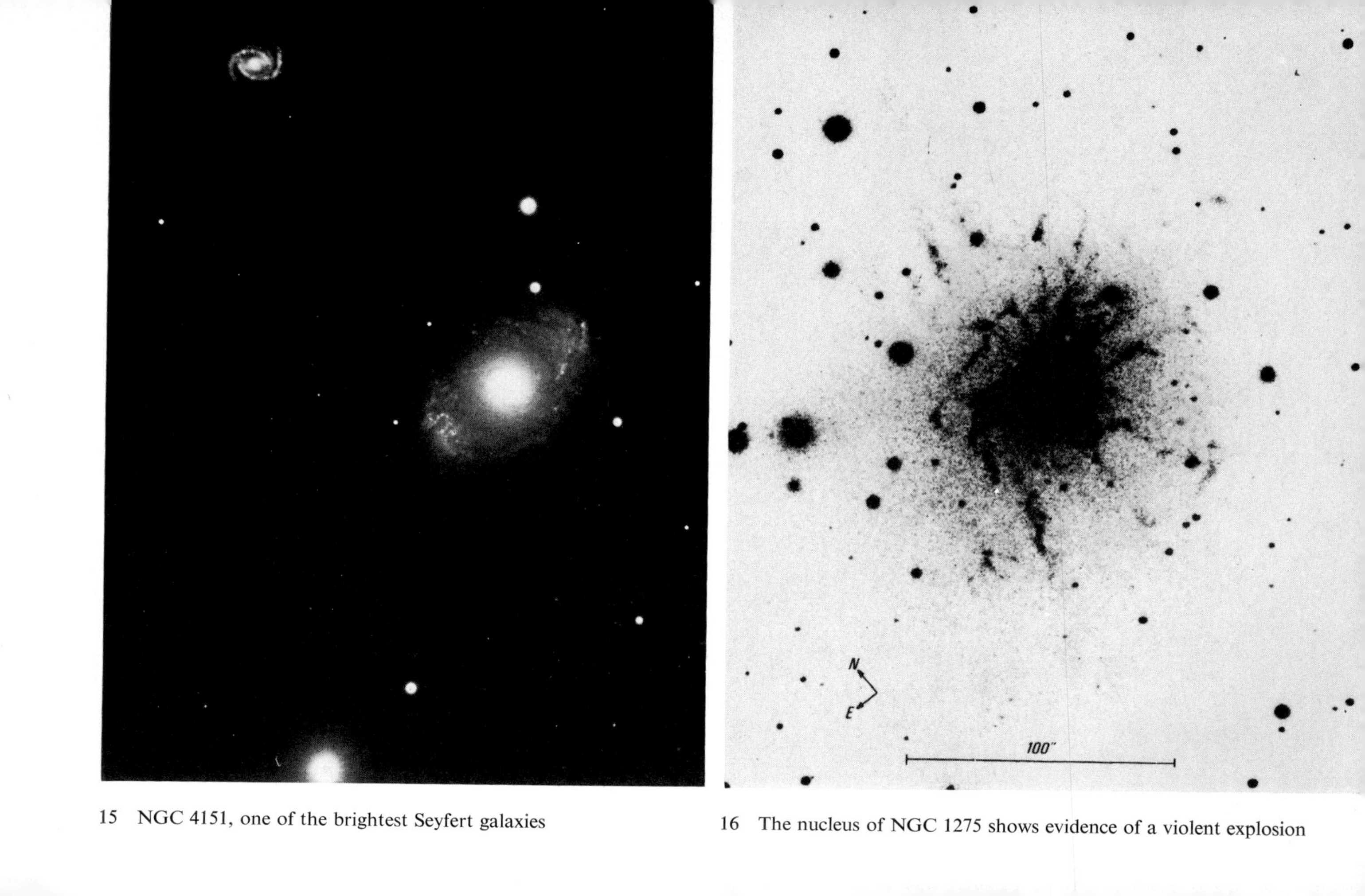

15 NGC 4151, one of the brightest Seyfert galaxies

16 The nucleus of NGC 1275 shows evidence of a violent explosion

17 A cluster of galaxies in Hydra

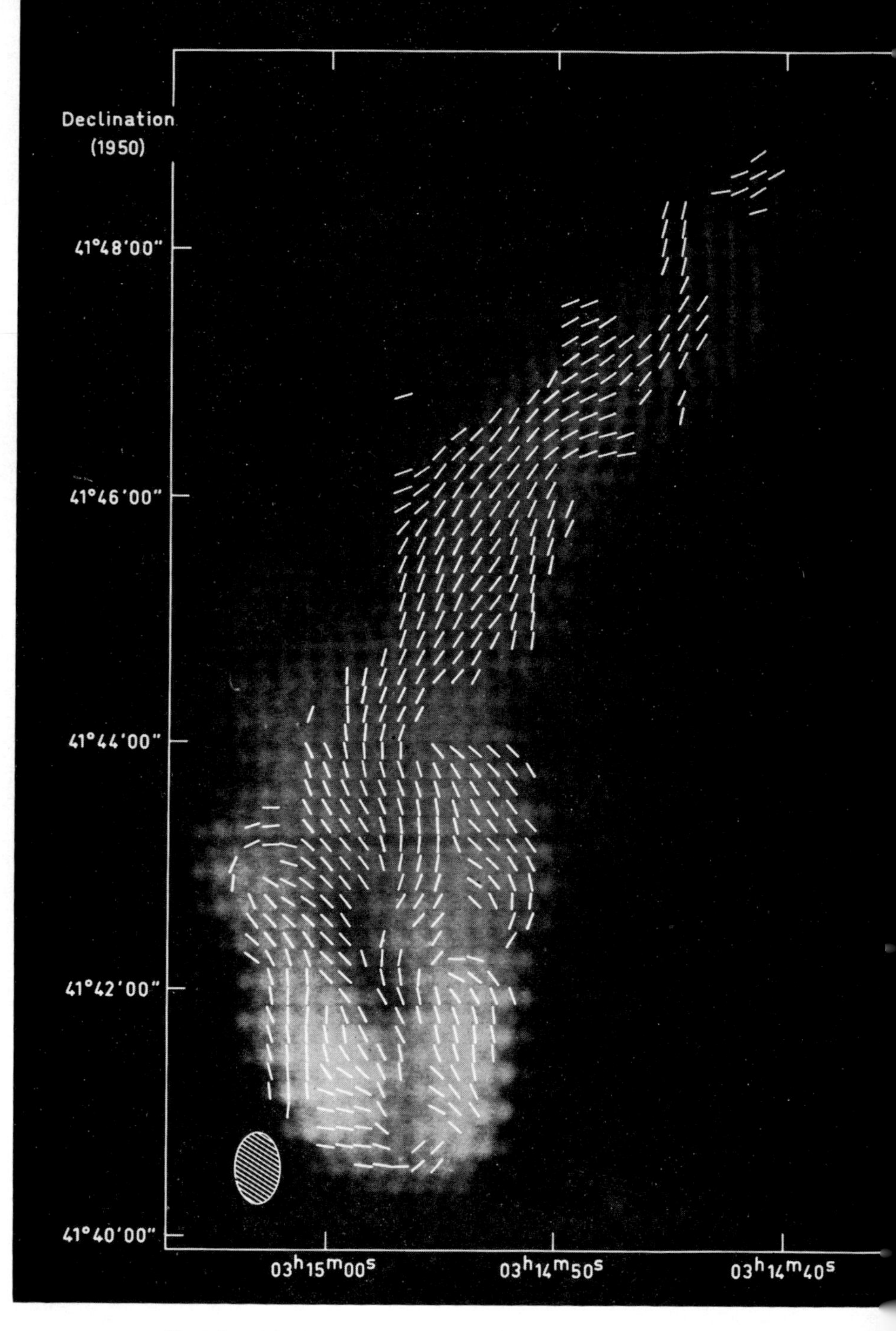

18 The white bars delineate the magnetic structure in the NGC 1265 tadpole galaxy

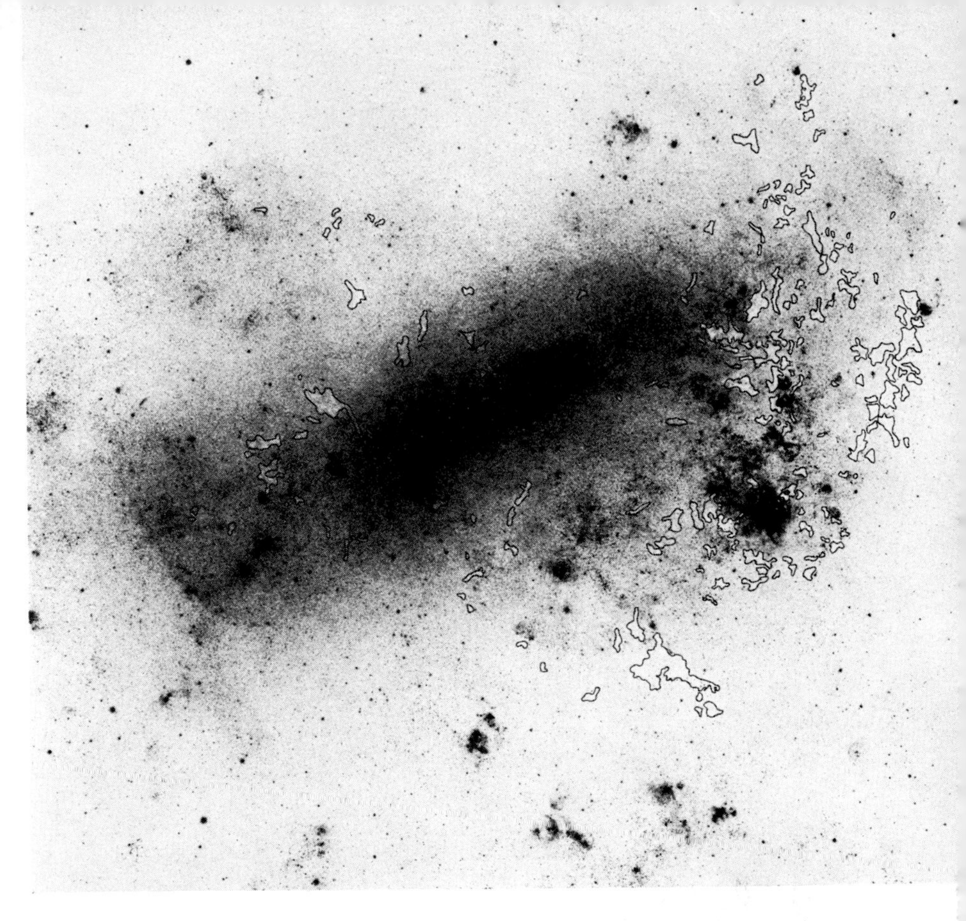

19 Dust clouds in the Small Magellanic Cloud are indicated by the black outlining

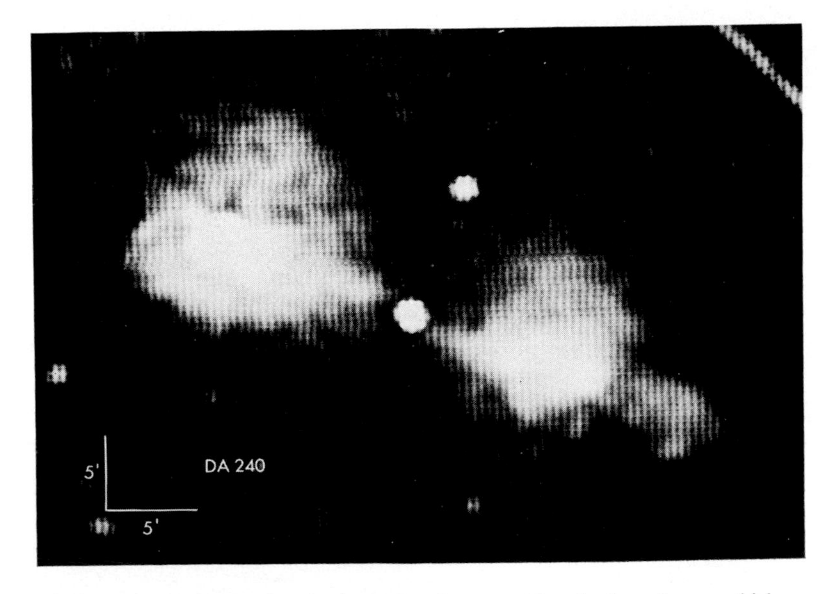

20 A radiograph of DA 240, the largest known object in the universe, which is six million light-years in extent

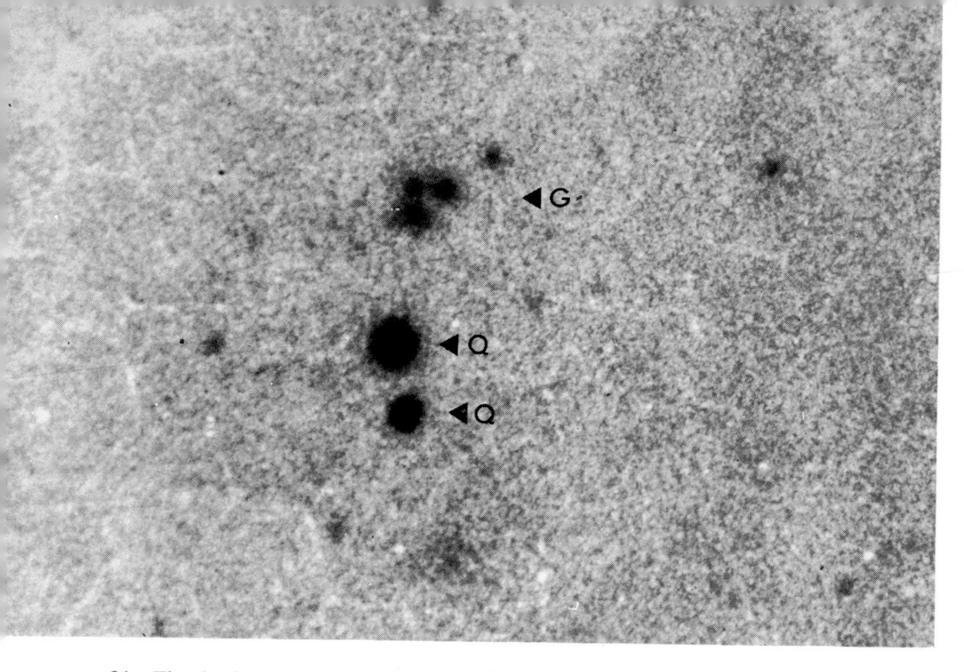

21 The double quasar 4C 11.50. This 78-minute exposure was taken with the telescope shown in Plate 1. Quasars indicated by **Q**, galaxies by G

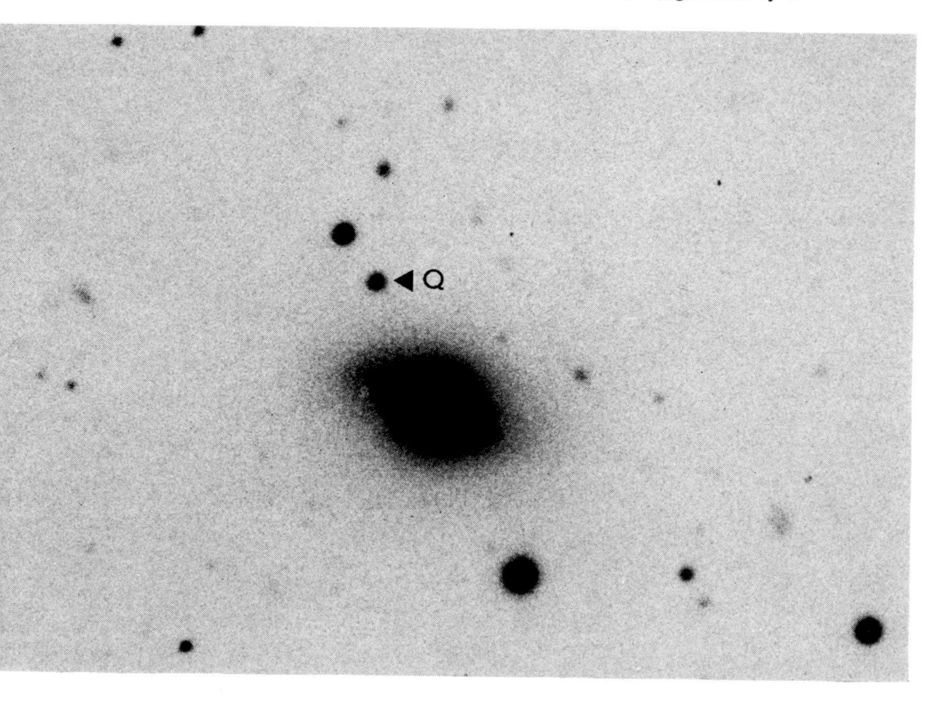

22 Is quasar PHL 1226 associated with the nearby galaxy?

during that time the central nuclear disc, which survived the paroxysms, shed up to 10^7 solar masses of material. The ejecta now form the high-velocity gas seen above and below the nuclear disc, and also the expanding arm of gas at 3 kpc. All the velocities have now been slowed down to their present values through collisions with the quiescent gas which was outside the nucleus before the explosive phase began. This generally tumultuous scenario receives further support from the presence of a strong radio source in the nucleus (see Figure 24) as well as the intense infrared source.

Even if our own galactic nucleus did not show any evidence of violent activity in the past, it would still be remarkable for the vast selection of molecules which have been discovered in its Stygian gas clouds since 1969. By observing emission and absorption lines in the microwave region of the spectrum, astronomers have now identified over two dozen species of molecules in the gas clouds of the Milky Way. Many of the molecules are organic substances, which

TABLE 6

Molecules found in interstellar clouds

Molecule	*Symbol*	*Typical location*
Methyl alcohol	CH_3OH	galactic centre, Sgr A
Acetyldehyde	$HCOCH_3$	galactic centre, Sgr A
Formamide	$HCONH_2$	galactic centre, Sgr B2
Hydroxyl	OH	compact H II regions
CH-radical	CH	supernova remnants, Cas A
Formaldehyde	H_2CO	galactic centre, Sgr B2
Methanimine	H_2CNH	galactic centre, Sgr B2
Cyanoacetylene	HC_3N	galactic centre, Sgr B2
Isocyanic acid	HNCO	galactic centre, Sgr B2
Water	H_2O	compact H II regions
Ammonia	NH_3	galactic centre, Sgr B2
Methyl alcohol	CH_3OH	Orion nebula
Carbon monosulphide	CS	galactic centre, Sgr B2
Hydrogen cyanide	HCN	Orion nebula
Silicon monoxide	SiO	Orion nebula
Methyl cyanide	CH_3CN	galactic centre, Sgr B2
Carbon monoxide	CO	Orion nebula

means that they contain the element carbon, and some of them are surprisingly complicated—at least in an astronomical context. For example, formic acid, acetyldehyde, methyl cyanide, dimethyl ether and methyl alcohol have all been tracked down in the nuclear clouds (Table 6). Almost all the known interstellar molecules have been seen in considerable concentrations close to the galactic centre. By measuring the precise emission or absorption frequency of a particular line it is possible to work out the velocity of discrete components of the galactic clouds. The profile of the spectral line and its central intensity yield further information on cloud temperatures and the density of molecules. Ammonia is a particularly useful probe of the molecular clouds because it has a rich spectrum of transitions. By comparing the intensities of certain transitions in this substance a picture of the velocity and temperature structure inside clouds can be built up.

There are two especially notable features of the molecular clouds at the hub of the Milky Way. One large radio source, Sagittarius A, appears to be at the galactic centre itself; another source of radio continuum radiation, Sagittarius B2, is to one side of Sgr A, but appears to be closely related to it. In Sgr B2 the particle density may rise as high as 10^{12} atoms m^{-3}, giving a total mass of 10^7 solar masses; to date, this is the biggest single conglomeration of molecular matter found in the Galaxy. Its central core is almost certainly unstable to the disruptive forces of its own gravity, and the object could therefore be collapsing on a time scale of 50,000–100,000 years. Is this the prelude to another giant galactic explosion of the type that van der Kruit has already postulated? Further mapping of the cloud will explore this view.

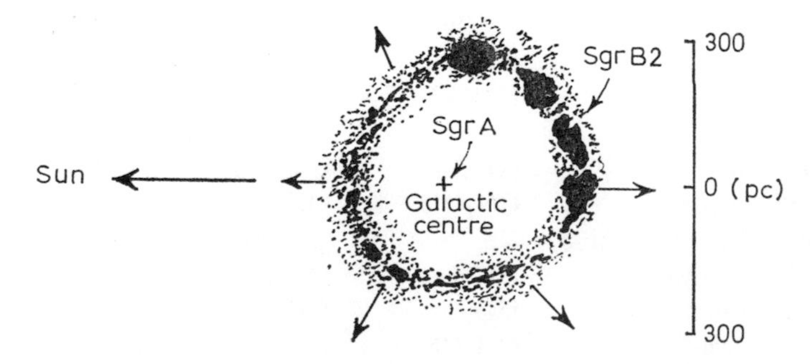

Fig. 25. A possible model of the molecular clouds in the galactic nucleus

Nicholas Scoville of Columbia University has constructed an intriguing picture of the traffic jam in the galactic nucleus (Figure 25). He finds that most of the gas at or close to the centre is probably concentrated into discrete clumps up to 30 pc in diameter, each containing roughly 1 million solar masses of gas. Individual clouds are moving at speeds in excess of 40 km/sec. Measurements of Doppler shifts in the radio lines indicate that the clouds are not participating in the normal pattern of galactic rotation; apparently, the matter near the nucleus is a law unto itself as far as motion is concerned. Scoville has postulated that the observations of molecular lines can be satisfactorily explained by a rotating ring of clouds, with a diameter of 300 pc, which is centred on the galactic nucleus and is expanding away from it. This great swirling smoke ring weighs in at 3×10^6 solar masses. Where did it all come from? A possible answer to this question comes from tracing the motion back in time: the velocity pattern indicates that the salvo of molecular clouds burst out of the nucleus in a detonation which occurred some 1 million years ago.

In summary then, we can say that dense and massive clouds, rich in organic molecules, appear to have been born in the galactic nucleus and they are now expanding away from the central region.

Another feature of our galactic nucleus which has caused much interest in recent years is the possibility that it may emit gravitational radiation. This is completely different in character to electromagnetic radiation, examples of which are radio, light and heat radiation, because it is associated with the force of gravity rather than vibrations of the electromagnetic field. Several groups of experimental physicists have attempted to detect vibrations due to gravitational waves from deep space. This is extremely difficult to do because the apparatus must somehow be isolated from all extraneous vibrations and oscillations, due to the passing traffic, the tides and so forth. The man who appears to have been the most successful is Joe Weber of the University of Maryland. He claims to have detected pulses of gravitational radiation from the heavens, with a maximum of signal strength directed towards the galactic centre. Other scientists have not confirmed Weber's claims, but it must be said that no one has tried them with exactly the same type of apparatus, or with Weber's tenacious determination. Some physicists are therefore sceptical that gravity waves from the nucleus have in fact been detected. Nonetheless, this has not led to a moratorium on theorizing.

If Weber has indeed detected gravity waves from the nucleus, then the implications for astrophysics are quite fantastic. In order to get the intensity of radiation which his apparatus is indicating, the galactic nucleus would have to destroy matter, and turn it into gravitational radiation energy, at a rate of hundreds of solar masses a year. This annihilation implies a continuous mass loss from the centre that would have profound effects on the evolution of the Galaxy, unless new matter is continuously created in the centre. The implications have not really been worked through in much detail, however; most theorists probably prefer to wait for convincing evidence of the radiation before committing themselves to long research programmes.

The overall picture of the centre of our Galaxy which emerges is a scene of considerable activity. Explosions of significant consequence would appear to have taken place in the recent past to judge by the large quantities of matter seen flowing away from the central regions. Although parts of the general picture are speculative, it is evident that our Galaxy is similar in its nuclear properties to active galaxies, although the actual scale of operation is lower than in the more spectacular cases. Nonetheless, it is very useful to have an example of a galactic explosion so near at hand, since by studying it in comparative detail we may learn more about the profligate galactic nuclei.

7.3 The nuclei of normal galaxies

In investigations of the nuclei of normal galaxies, standard optical techniques can be used, since we are not normally shielded from the centres of other galaxies by great swathes of dust. The important investigations which can be made are of the type of stars present in the nucleus, the state that the gaseous component is in, and the structure and dynamics of both components. Many of the normal giant galaxies possess brightness concentrations at their centre. In M31 there is a sharp spike of apparent magnitude 12 mag standing out clearly against the faint background. Its angular size of 1.6×2.8 arc sec corresponds to a linear size of 5.4×9.4 pc (diameter). Like many nuclei the one in M31 contains ionized gas. Spectroscopic surveys show emission lines from oxygen atoms that have been stripped of one electron. This emission line, at 3727 Å, is a prime indicator of the presence of gas. Other indicators are the red light of hydrogen Hα at 6563 Å, and lines of singly ionized nitrogen at 6583 and 6548 Å. In M31 the gas is blowing gently out of the centre and this gives a mass loss of 0.01 solar masses per year. Gas is present in the nuclei of the well-developed spirals, but decreases in proportion as one goes along the Hubble sequence of galactic types from Sc through to Sa. It is usually difficult to put a precise value on the amount of gas actually present, because the neutral atoms do not give rise to strong emission lines, but it is generally reckoned that normal nuclei contain up to 10^6 solar masses radiating the observed emission lines. This value is small compared to the total mass of the nucleus.

For the nuclear gas to radiate, energy must be fed into it. In some systems many hot O and B stars give out enough energy to excite the gas and cause loss of electrons, or ionization. In elliptical galaxies and spirals such as M51 and M81 the energy to heat the gas may be released during collisions of stars.

Stars in the nucleus are investigated by measuring colours and obtaining spectra of the central regions. Galaxies such as M31 show the characteristics of old, well-evolved stars in their nuclear spectra, and consequently it has been inferred that the nucleus is mainly inhabited by giant stars that are well evolved and comparatively cool, that is, less than 5500 K surface temperatures. Investigations of the star density in the nucleus of M31 yield a density of 10^5 stars per cubic parsec. The average spacing between the stars packed in this tightly is only 2000 astronomical units. Even so, the probability of one star colliding with a neighbour is very small; perhaps one crash occurs every million years or so on average in M31. Some nuclei are denser and then the collision rate is, of course, higher.

We have already seen that the nucleus of our Galaxy is dynamically very interesting, because it has evidently suffered from considerable explosions. The motion of stars and gas in other nuclei can be derived from shifts in the spectral lines. In M31 it has been found that the nucleus is spinning with a rotational velocity which attains a maximum value of 87 km/sec at 6 pc from the centre. From this dynamical information a mass estimate of 1.3×10^7 $M_\odot$ is derived from the virial theorem. The gas in M31 shows major departures from uniform motion, and these can best be explained in terms of an expansion away from the centre. Up to one solar mass may be shed each year from the nucleus. A satellite companion to M31, the dwarf elliptical M32, also displays nuclear rotation which is consistent with a central mass of 10^7 $M_\odot$. Merle Walker has concluded that this nucleus may be dynamically separate from the rest of M32 on account of its rapid rotation.

The nearest Sc spiral is M33, and its nucleus can be examined in some detail. This nucleus mainly comprises a compact cluster of blue stars (spectral types A5–A7) with a diameter of 17 pc. No substantial rotation is evident. Another nearby spiral is M51, which is characterized by a complex nuclear region. Within 300 pc of the centre is a multitude of bright H II regions, along with some wispy dust lanes. Up to 10^6 $M_\odot$ of excited gas is present in the nucleus, a value that indicates the presence of an underlying energy source. Large numbers of very hot (400,000 K) stars or an ultraviolet emission source could be feeding energy into the hydrogen and heating up the H II regions. Intense radiation from the gas may be masking the light from a population of dwarf stars in the nucleus. The Burbidges have shown that the gas is moving in complex, non-circular patterns, with velocities of the order 100 km/sec. It is possible that fans of gas are now being expelled in many directions from the nucleus; in view of the large mass established for this gas, the kinetic energy of motion most be substantial. There is plenty of evidence that the nucleus of M51 is in a state of energetic turmoil.

In the southern hemisphere the barred spirals NGC 1097 and 1365 merit attention. Both nuclei contain giant H II regions, and the emission lines of hydrogen and nitrogen show that substantial non-circular motion is present. A combination of expansion and rotation is implied by the spectra of the nuclei.

7.4 Seyfert galaxies

Carl Seyfert, in 1943, drew attention to a small class of spiral galaxies whose nuclei set them apart from other systems known at the time. He found that a small proportion, about 1 per cent, of spirals have inconspicuous arms and extremely bright central regions. He gave a list of twelve such systems, having starlike nuclei, highly excited spectra and exceedingly broad emission lines. These are the characteristic properties of Seyfert galaxies. Their nuclei are currently the subject of much topical research, as they are scrutinized with optical, radio, infrared and X-ray telescopes. Radiation from the nucleus is entirely due to hot gas, within which violent motion can be discerned. Dozens of objects are now in the Seyfert category, the list having been much extended by the efforts of Wallace Sargent. The brightest, and consequently the most thoroughly studied, are NGC 1068, 1275 and 4151; their redshifts are 0.0036, 0.018 and 0.0033 respectively. Optical photographs reveal nuclei a few arc seconds across, corresponding to linear diameters of a few hundred parsecs (Plate 15).

High resolution optical spectra of Seyfert nuclei show the conspicuous feature first noted by Seyfert, namely the presence of bright lines of emission from highly excited gas; neon, oxygen, sulphur, iron, hydrogen and argon are common. Oxygen is sometimes found in three states of ionization: neutral oxygen, singly ionized with one electron removed, and doubly ionized with two electrons missing; this indicates a wide range of excitation conditions. In NGC 4151 there are lines of [Fe X] and [Fe XIV], that is iron atoms with nine and thirteen electrons absent, which also appear in the high-temperature corona of the sun. Highly excited atoms are relatively so common in Seyfert nuclei that the emission radiation must be coming from gas clouds rather than the coronal regions of hot stars. Indeed, for NGC 4151, which has the richest emission spectrum, there is no direct evidence of a stellar population in the nucleus; 28 per cent of the radiation is concentrated in the emission lines.

Measurements of the relative intensities of certain emission lines in a Seyfert galaxy spectrum enable the estimation of the prevailing temperature and density. The relative strengths of lines in the species of oxygen are used to calculate both temperature and density. Electron densities, for example, can be derived independently from both oxygen and sulphur lines. It is impossible to make a single homogeneous model of the Seyfert nucleus, simply because the spectra are so rich. It is also important to realize that if there is an underlying energy source which may be injecting high-energy electrons into the nuclear clouds then the simple models will not give a complete picture. Nevertheless, the models at least indicate the likely range of physical conditions.

Among the best studied at the Seyfert galaxies is NGC 1068, for which the Burbidges and Kevin Prendergast estimated a total mass of 2.6×10^9 $M_\odot$ within 2 kpc of the centre. A detailed dynamical study of this zone also showed that any central condensation in the nucleus would not exceed 3×10^7 $M_\odot$. NGC 1068 would appear to have a less concentrated nucleus than most Seyfert galaxies. With an image-tube spectrograph Merle Walker has measured the profile of individual emission lines in NGC 1068 and he concludes that there are several distinct clouds moving with velocities up to 600 km/sec. Cloud diameters are put at 200–350 pc, their masses at 10^6–10^7 $M_\odot$, and their electron temperatures are around 20,000 K.

An interesting feature of NGC 1068 is the powerful flux in the infrared out to a wavelength of 100 μm. The total infrared luminosity is equivalent to 10^{11} solar luminosities, so that the greatest proportion of energy emission lies in the infrared. In 1972, G. H. Rieke and Frank Low of the University of Arizona reported the observations of sudden changes in the 10-μm emissivity; they saw the signal strength alter by factors of 2 or 3 in a matter of weeks. During one two-month session they spotted an alteration at 10 μm that amounted to a total change of 7×10^9 solar luminosities; all we need do to put this into perspective is note that the amount of power which the nucleus of NGC 1068 is switching on and off in a matter of weeks is comparable to the total luminosity of our Galaxy!

The variations in NGC 1068 require a non-thermal mechanism for the generating source of the intense infrared emission. Models in which dust is heated to the temperature at which it radiates strongly at 10 μm appear doomed to failure as there is no means of switching the dust radiators on and off rapidly. Because of difficulties with the hot-dust concept, Rieke and Low prefer to attribute the radiation to a mysterious non-thermal source. This means that the energy is released by a mechanism which does not involve heating matter; an example of such a mechanism would be synchrotron radiation, that is emission from electrons being accelerated in a magnetic field.

Another Seyfert of great interest is NGC 1275, which has a rich emission-line spectrum with two distinct velocity components. Immediately outside the nucleus gas is moving with a very high velocity of 3000 km/sec; these motions can be traced outwards for 10 kpc. The high-speed gas has an emission spectrum with sharp lines. In 1970 Roger Lynds obtained remarkable photographs (Plate 16) of NGC 1275, taken at Kitt Peak National Observatory by means of isolating filters centred on the apparent wavelength of the red light of hydrogen for the two velocity systems. These pictures have resolved tangled fingers of gas travelling at different velocities. They show the enormous extent of the filamentary structure which appears to have been generated in a massive explosion, and they confirm the velocity components derived spectroscopically. NGC 1275 superficially resembles the twisted tortuous structure of the Crab Nebula but on a vastly greater scale. The continuum light from NGC 1275 conveys further evidence for unusual events because it rises steeply towards the infrared. The distribution of intensity through the spectrum follows the power law pattern so characteristic of non-thermal radiation. Furthermore, NGC 1275 is the dominant member of the Perseus cluster of galaxies and a strong radio source.

Of all the Seyfert nuclei NGC 4151 is the best studied. So brilliant is its nucleus that no outer galactic structure would be distinguishable if it were very much further away than it is. It has the richest spectrum, the family of lines running up to the exotic coronal lines [Fe X] and [Fe XIV]. From an analysis of the line intensities John Oke and Wallace Sargent concluded that the emission lines from a region with an electron density of 5×10^9 m^{-3}—astonishingly high for gaseous material in galaxies—had a temperature of 20,000 K. The total mass of the gas nebulae and filaments in the nucleus is 2×10^5 $\bar{M}$. Kurt Anderson and Robert Kraft isolated three velocity components in NGC 4151 and they concluded that three giant shells of gas, with velocities of 280, 550 and 840 km/sec, have been blasted out of the nucleus. Mass loss rates of 10 to 1000 solar masses per year during the sporadic bursts of activity are indicated. The light from NGC 4151 is variable on a time scale of a year or so.

Hydrogen emission lines in the nuclei of the Seyferts pose perplexing problems on account of their great width. The hydrogen lines consist of an intense central core that has a width similar to the oxygen and nitrogen lines, and, spread outside the core, two very broad wings. Exactly how the enormous widths of the hydrogen lines arise is still an open question. If the extension of the line profile is attributed to velocity effects large dispersions are involved. A few examples illustrate this: in NGC 566 the line spread expressed in terms of velocity is 3500 km/sec, NGC 4051 gives 3600 km/sec and NGC 4151 an impressive 7500 km/sec.

There are several intriguing ways in which the large widths could arise. Possibly the velocities are true speeds of bulk motion for the hydrogen gas, in which case an extremely massive, highly condensed, nuclear core would be necessary for anchoring the swiftly moving gas down at the central hub. A second possibility is that the hydrogen gas is in a state of tremendous excitation and turmoil, so that the spread in the random velocities of individual atoms runs into thousands of kilometres per second. Such a situation could arise in explosive circumstances. Random motions of atoms in a gas always broaden the spectral lines, so that large line widths can be taken to indicate great agitation. Yet another suggestion is that photons from the hydrogen are modified in frequency as they pass through a hot ionized gas. Ionization releases electrons, and if these move rapidly and collide with photons, then energy is exchanged. The energy change is essentially statistical so that the energy spectrum of the photons becomes blurred. Therefore the emission can start out as a narrow line, but becomes broadened if it has to traverse a hot gas. The process of electron scattering has most influence on the hydrogen frequencies under astrophysical conditions. So only the hydrogen will have the broad wings. Much more research needs to be conducted on Seyfert nuclei in order to understand more fully the enigmatic line-broadening processes.

Among the less well-observed Seyferts there are several objects worthy of mention. Vera Rubin and Kent Ford have discovered discrete clouds speeding along at thousands of kilometres per second in NGC 3227. There appears to be a whole cloud system expanding outwards at 150 km/sec. Matter is probably being continuously ejected from the nucleus. NGC 3516 is a bit of a puzzle too. The line spectrum and continuum have varied substantially in the last thirty years; lines that Seyfert saw have disappeared and new ones have popped up instead. Astrophysicists have tentatively attributed this change to the sudden ejection, from a central source, of dense clouds (100 solar masses) at around 1000 km/sec. Among the galaxies categorized since Seyfert's classical paper is 3C 120, a radio source associated with a compact galaxy. The object has shown rapid variations at optical wavelengths, amounting to factors of two in only months, and the radio emission has undergone a series of successive outbursts.

Intense infrared emission is present in several Seyfert nuclei, for example, NGC 1068, 3227, 4051, 4151 and 3C 120. The most important aspect of this work is that the derived infrared luminosities amount to 10^{39} watts. In these objects the energy loss at infrared wavelengths is one or two orders of magnitude greater than at optical or radio wavelengths. Short-term variations of around a magnitude are also present, completing a picture of Seyferts in which fluctuations can be seen right through the electromagnetic spectrum. The energy loss rate is equivalent to the total destruction of 0.5 solar masses per year in NGC 1068. No matter how the energy is produced the nuclear mass must be decreasing at this rate in order to keep up the power output. We do not know how long NGC 1068 has performed at its present level; in 10^8 years the mass in the nucleus would utterly vanish if the present luminosity continued. It is clear that the process is either sporadic or short-lived, or else the mass is continuously replenished.

Finally under the heading of Seyfert galaxies, one other class of galaxy should be mentioned—the N–galaxies. These were originally defined as systems with brilliant nuclei superimposed on a considerably fainter background. As far as form is concerned, they are essentially identical to Seyferts. However, the majority of N–galaxies were so-called by radio astronomers who identified them with strong radio sources. Furthermore they do not have quite such broad lines as Seyfert nuclei. Some (3C 109, 371 and 390.3) are optically variable, indicating the presence of small components with dimensions of only a few light years.

7.5 Nuclei as galactic power-stations

Observers have now catalogued several key properties of galactic nuclei. A large fraction of the nuclei have some degree of activity, and matter is ejected from nuclei, often in the form of massive gas clouds and fast-moving atomic particles. Even in conventional galaxies the density of stars becomes very high at the nucleus. The more exotic nuclei frequently contain a small radio source, a further signature of unusual conditions. Often the infrared flux is overwhelmingly vast and poses severe theoretical problems. Physical conditions in the nuclei involve dense, hot, regions of gas, together with an energy source capable of ionizing the gas. Variability in the power shows that at least some of the active regions are very small. It is impossible, without invoking special geometry, to switch a source of energy on and off faster than the time taken for light to cross that energy source. This means that the time taken for substantial changes to manifest themselves can be used as an upper limit on the size of the energy source. It should be borne in mind that variability at some level is the rule rather than the exception in most powerful galactic nuclei. Hence we reach the inescapable conclusion that the mysterious power-stations in the nuclei are exceedingly small, pouring out 10^{36}–10^{39} watts from regions only a matter of light years in diameter. Fluctuations in the infrared luminosities of NGC 1068 and 4151 appear to rule out models in which the far infrared energy flux is produced by heating up specks of dust!

In 1971 Geoffrey Burbidge stated that an understanding of galactic nuclei is of the greatest importance to extragalactic astronomy. We are still only beginning as far as the observations and theory are concerned. In future years the new generation of large optical telescopes and the further improvement of infrared, radio and X-ray techniques should permit more rapid advance. These studies are vital if we are ever to get to grips with understanding the formation and evolution of galaxies and the overall picture of energy generation in a violent universe.

8 · Clusters of galaxies

8.1 Families and hierarchies of galaxies

Edwin Hubble determined the basic rules governing galactic family trees in the 1920s and early 1930s from surveys of the density of galaxies in sample fields which he photographed with the 1.5-metre and 2.5-metre telescopes on Mount Wilson. Although the large-scale structure of the universe appears to be homogeneous and isotropic, there is also local structure of great complexity. What Hubble discovered was that a general background distribution of galaxies was broken up by large clusters of galaxies a few megaparsecs in size. With a small telescope the brightest galaxies in a great cluster in Coma Berenices can be seen quite easily; Coma has about 800 members and is among the nearest of the great clusters. The Hydra cluster is shown in Plate 17.

Observers who continued Hubble's work soon realized that the concept of a uniform sea of galaxies disturbed by the occasional cluster was not correct. Wide-field astrographs of the Harvard College Observatory and the Palomar 45-cm Schmidt camera showed that clumps and clusters are the rule rather than the exception. E. Holmberg estimated in 1974 that 70 per cent of all galaxies are in a group dominated by a giant galaxy. Clustering appears to be a fundamental property of the distribution of galaxies in space. In short, galaxies have a strong preference for living in families.

The Palomar Sky Survey has provided plenty of information on galactic hierarchies. Clustering occurs on scales ranging from very small aggregates, such as Stephan's quintet, right up to enormous superclusters 50 Mpc across. Within larger clusters it is often possible to distinguish sub-clustering. In our locale, for example, there are two well-defined sub-groups, one centred on the Milky Way and the other on M31. In turn it is conceivable that the Local Group and the

Virgo cluster (11 Mpc away) are components of a local supercluster. According to George Abell, the gathering of broods of clusters into superclusters is a fairly general occurrence. This raises the question as to whether the hierarchy of clustering continues ever upwards to more and more giant structures. In 1958 Abell concluded that the maximum scale for clustering is around 60 Mpc; this confirmed conclusions reached earlier by Edwin Hubble and Fritz Zwicky. The existence of an upper boundary to the clustering phenomenon is supported by other observations that show that the grand architecture of the universe is homogeneous and isotropic. Truly immense superclusters would destroy the observed overall symmetry and uniformity of the universe.

8.2 Our own family of galaxies

The galaxy cluster to which the Milky Way belongs is known as the Local Group, and it consists of three associated clumps of galaxies (Figure 26). The Milky Way, Magellanic Clouds and a handful of dwarf systems comprise one condensation. About 600 kpc away is a second concentration composed of M31, M33 and their dwarf companions. Finally, a third multiplet may be formed by the galaxies Maffei 1 and 2, discovered in 1970 close to the plane of the Milky Way, along with IC 10 and IC 342. The overall extent of the Local Group is about 2 Mpc, with the sub-groups spanning some 500 kpc each. Including the dwarf galaxies there are about 200 known members of the Local Group.

It is fortunate that the Milky Way should be part of a cluster of galaxies. Within the Local Group there are available for our study several different types of galaxy, each of which can be investigated comparatively well because of its close proximity. It is especially important that the properties of the populations of stars can be studied in detail so that the influence of stellar populations on galactic evolution and morphology can be derived observationally. The different galactic types are represented well: M31 and M33 are spirals of type Sb and Sc. M31 is accompanied by the E2 galaxy M32 and a dwarf E0. Irregularity is the keynote of the Large and Small Magellanic Clouds, although it is possible that the LMC is a barred spiral. One type of sidereal organization, that of dwarf spheroidal galaxies, can only be investigated properly within the Local Group.

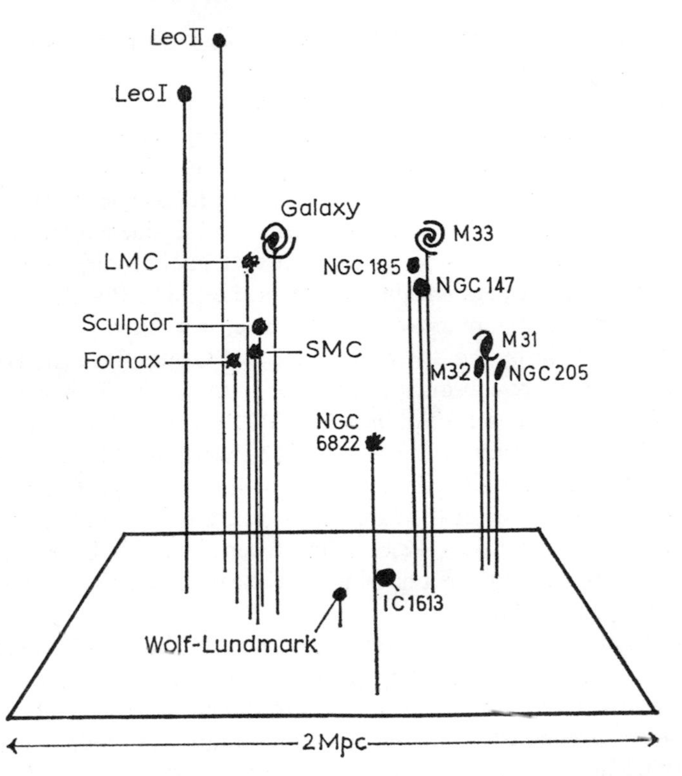

Fig. 26. Galaxies of the Local Group form three sub-clusters: M31, M33, and our Galaxy

An important discovery was made by Harlow Shapley in 1937 when a galaxy within the Local Group, of an apparently unknown type was found on plates exposed at the Harvard Observatory's Boyden Station in South Africa. In the Sculptor constellation Shapley had found an assemblage of stars unlike any previously described; it consisted of a very diffuse and extensive aggregate of extremely faint stars. On the discovery plates the stars appeared to be of roughly the same faint magnitude, close to the plate limit in fact, and they formed a uniform distribution with a circular outline. Astronomers at Harvard College and Mount Wilson subsequently demonstrated that this Sculptor system was a galaxy of stars, every one of which could be resolved, because the star density within the galaxy was so low. In 1938 Shapley found another example in

Fornax, and in the 1950s the Leo I, Leo II, Ursa Major and Draco dwarf galaxies were discovered in northern skies.

Perhaps the best way of describing the dwarf galaxies is as 'super globular clusters' of stars that are extremely underluminous. For example, the absolute magnitude of Leo I is —11 magnitudes, so that the entire galaxy is less energetic than the single brightest star in the Magellanic Clouds! Even so these feeble sidereal families are true galaxies, as witnessed by the fact that Paul Hodge located five faint globular clusters within Fornax in the 1960s. But they contain no gas or dust as far as we can perceive, just stars.

Morphologically the dwarfs in the Local Group form the tail-end of a sequence that commences with the giant ellipticals such as M87, passes through dwarf ellipticals like M31's satellites and continues on down to Sculptor and Fornax. The dwarfs mentioned here are literally within a stone's throw of the Milky Way, all of them being within 300 kpc. One of the faintest members is in Draco, and it has an absolute magnitude —8.6. Almost certainly this lower limit is imposed by our instruments which are incapable of finding galaxies any dimmer than this.

Since at least six dwarf systems exist close to the Milky Way, there is raised at once the exciting possibility that most galaxies in the universe may be subluminous, and are therefore undetectable. This view was strengthened in 1972 by the observations of Sidney van den Bergh, who has found four new faint systems within the Local Group region. He realized that galaxies similar to Fornax and Sculptor would just be detectable at the distance of M31 with the 1.2-m Schmidt camera at Mount Palomar. Inspired by this he commenced a search of the sky near to the Andromeda Nebula (M31) using plates with fast emulsion. These sensitive plates showed faint smudges of light, with very little central condensation, in the vicinity of Andromeda. Van den Bergh has called these smudges And I, II, III and IV, and identified them as remote dwarf spheroidal galaxies. Each member of this entourage probably contains only a few hundred thousand very old stars. In 1973 further work on And III, namely photography with the Hale Observatories' 5-metre telescope, resolved it into yellow stars of magnitude 22.2 and fainter. Although no galactic nucleus is present, the star density increases towards the centre as we expect for a galaxy. The linear size of And III is 600 by 900 parsecs and its total absolute magnitude is —11, the same as that of a rich globular cluster of stars.

Studies of the colours and magnitudes of stars in the dwarfs near the Milky Way give results similar to the globular clusters. The implication is that there are no significant numbers of young stars in the dwarfs, a property which immediately distinguishes them from larger spiral galaxies. Numbers of variable stars range from 100 RR Lyrae-type in the Ursa Major galaxy, up to 700 in Sculptor.

Two further members of the Local Group are NGC 6822, at 460 kpc from the Milky Way, and IC 1613, which is 740 kpc away. These are both irregular dwarfs. NGC 6822 is in the constellation Sagittarius; it was picked up first by E. E. Barnard in 1884 at Nashville, Tennessee. When the great telescopes at Mount Wilson came into action it was among the first galaxies to be studied in great detail. Edwin Hubble showed, in 1925, that it must be beyond our Galaxy by determining its physical size and distance from his observations of eleven Cepheid variable stars. The galaxy is 2.7 kpc long in maximum extent. It contains a number of bright gas clouds, and the chemical composition of these has been investigated; this has shown that the galaxy is deficient in heavier elements, such as oxygen and nitrogen, in comparison with our Galaxy. Such differences may indicate that the processes for manufacturing heavy elements differ between giant and dwarf galaxies.

The dwarf irregular IC 1613 is 5 kpc across. It contains a whole variety of star types, which means that it is a useful test bed for checking our ideas of the pattern of stellar evolution in galaxies. Different groups of stars within the IC 1613 galaxy have been investigated and the age-structure of the galaxy deduced from studies of the Cepheids. The dwarf is simple and uncomplicated; furthermore it has not got any star clusters. This sobering fact highlights the difficulty faced by theorists who want to spin a galactic scenario of star clusters; nobody knows why some giant galaxies have few clusters and others possess many; we do not understand what it is that determines the parameters of star cluster formation.

So far we have only discussed the local midgets. But the most conspicuous members of the Local Group are those in the Andromeda system, composed of M31, M33 and their companions. A great deal is known about M31 and M33 from optical and radio studies, and consequently they are both frequently mentioned in this book. M31 has a composite structure that is typical of many giant spirals. The central region resembles a giant globular cluster in many respects, having a smoothly changing luminosity gradient, although there is a bright nucleus. In the disc two splendid spiral arms twist their way from the centre to the periphery. Luminous stars, cloud complexes, dust lanes and hydrogen show up prominently. It is perhaps unfortunate that M31 is so steeply inclined to the line of sight, being only 12° away from the edge-on configuration. This has made it difficult to unravel the arm structure optically because dust in the near edge veils the more distant structures. However, radio astronomers have come to the rescue with beautiful maps of the delicate spiral structures which they have obtained from 21-cm measurements of the distribution of neutral hydrogen (Plate 6). Andromeda is very massive. From an analysis of the motion of its globular clusters, F. Hartwick and Wallace Sargent have deduced a mass of 3.4×10^{11} solar masses. Andromeda is one of the very few normal galaxies from which X-rays have been detected. In 1974 a group at the University of California described measurements obtained with a rocket-borne counter. They discovered an X-ray flux from M31 of about 10^{32} watts in the 0.5–5 keV band. The bulk of this energy is probably whipped up by small bands of highly exotic stars, similar to the highly luminous X-ray sources in our Galaxy.

Two bright minions, NGC 205 and M32, attend Andromeda. NGC 205 is the elongated one, and it orbits so close that it swims through the outer star fields of M31. In fact it is not possible to tell where the stars of one galaxy end and those of the next begin! M32 is very much like an oversize globular cluster. It has 4×10^{9} solar masses of stars with a composition similar to the globular cluster stars.

The other nearby giant is M33, a fantastically beautiful open-armed spiral that is virtually face-on. Colour pictures show it to be resplendent in brilliant blue and red stars, highly excited gas clouds, and murky dust lanes. It is about the same distance as M31, but only around one-tenth as massive. Radio maps of M33 are disappointing because they do not show the expected spiral structures. The galaxy must be relatively deficient in neutral hydrogen gas.

8.3 The nearest great clusters

As we conduct a census of the inhabitants of surrounding neighbourhoods, the next thickly populated district after the Local Group is the impressive Virgo cluster. This collection of 8-mag nebulae is centred on right ascension 12^h 30^m, declination $+12°$, in a triangle defined by Arcturus, Regulus and Spica. Its brighter members should be visible in a 15-centimetre or 20-centimetre amateur telescope. Extending north of the constellation Virgo, and on into Coma Berenices, are the loose suburbs of the cluster.

The Virgo cluster is around 11 Mpc from the Galaxy. Spiral galaxies constitute 75 per cent of the brighter members, the remainder being ellipticals along with a sprinkling of irregulars. Fainter members of the family include larger numbers of irregulars. The recession velocity for the Virgo group as a whole is about 1100 km/sec, but there is much activity within the cluster. Individuals are dashing around at up to 2000 km/sec relative to each other. Assuming that the cluster is stabilized by gravity, we can compute the amount of mass which is required to restrain the seething turmoil. It comes to an incredible 2×10^{11} solar masses per member, which seems to be implausibly large. Several explanations have been explored: does the cluster contain most of its matter in a non-luminous form? Is it really as stable as we are assuming? Or are galactic masses truly greater than most astronomers have dared to believe? Only further intensive investigations can answer such matters.

Somewhat further than Virgo is the great globular cluster of galaxies in Coma, which is around 70 Mpc away. This has about 800 members and a total mass around 10^{15} solar masses. As in the case of Virgo the mass per galaxy is anomalously high. A remarkable property of the Coma cluster is its richness and homogeneity. Only a few per cent of all clusters have a hundred or more galaxies in the first few magnitudes, as Coma has. Coma has two radio galaxies at its heart: NGC 4869 and the supergiant 4874. The source associated with NGC 4869 is known as 5C 4.81; it is an example of an interacting radio galaxy, displaying the characteristic 'tadpole' structure (Chapter 9). In addition to the two compact radio sources, Coma also has an extended source nearly 1° in size, which may be due to emission from gas filling the spaces between galaxies.

8.4 X-rays from clusters

The most significant discovery in extragalactic astronomy made by telescopes in satellites has probably been the detection of X-ray emission from extended regions in clusters of galaxies. The Perseus, Virgo, Centaurus and Coma clusters all register X-rays. In the two clusters with active galaxies, Virgo and Perseus, the emission is centred where the action is. Cluster X-ray sources are much bigger than any of the member galaxies. They stretch over about 1 Mpc, and have X-ray luminosities of 10^{38} watts. All the signs are that these are truly diffuse objects and are not associated with any one galaxy in the cluster.

Theorists speculate that a hot thin gas threads rich clusters, and that this gives rise to the X-ray emission. If they are correct a temperature of 10^7–10^8 K is indicated. How can we get the gas this hot? An idea which is currently in favour is that the active nuclei, radio sources or interacting galaxies in the cluster could heat up the gas with their own energy efflux. Supporting this notion is the fact that X-ray luminosity correlates with how fast the galaxies in the cluster are racing round. The detection of diffuse emission from the clusters certainly furnishes vital clues on the physics of the intracluster gas, and may also yield cosmologically important facts because the mean density of the intergalactic space partially influences the future evolution of the universe.

9 · Radio galaxies

9.1 Discovery of dis[illegible] radio sources

In every branch of s[illegible]e philosophers and historians like to pin-point golden moment[illegible]en whole new vistas were opened up by a particularly significan[illegible]rn of events. For radio astronomers the discovery of radio gal[illegible]es in the early 1950s was such an event, for it transformed an inte[illegible]sting astronomical sideline into a discipline of immense importanc[illegible] Radio galaxies have puzzled theorists for over twenty years, be[illegible]use they are objects in which a sizeable fraction of a galactic ma[illegible] has been transmuted to an exotic form of energy called relativistic [illegible]lasma. They have also been a fascination to cosmologists becaus[illegible] their immense luminosities enable their radio emission to be detected even if they are beyond the reach of optical telescopes. So there is the possibility of studying galactic properties at very great distances, during earlier phases of the universe, and of determining the structure of the universe at large distances.

The detective work which led to the discovery of radio galaxies can be traced back to 1944, when the U.S. amateur astronomer, Grote Reber, found indications of a discrete source of cosmic radio waves in the constellation Cygnus. In 1946, J. S. Hey, S. J. Parsons and J. W. Phillips, working in England, announced that the source in Cygnus definitely existed, but with their telescope capturing all the cosmic radio waves streaming in down a beam-width of 12° they could not accurately fix its celestial location. So there was no hope of identifying the source, named Cygnus A, with an optical object. During the next few years a British team under Martin Ryle and an Australian group directed by John Bolton set about surveying the heavens in a search for more radio sources. Before long they had found about 100, although the positional uncertainties were of the order of one degree. Consequently only a handful of pathological objects was matched up to optical counterparts: Taurus A with the Crab Nebula, and Centaurus A with NGC 5128, for example. Until 1954 some radio astronomers considered that most of the emission came from radio stars in our Galaxy.

Improvements in radio techniques led to a momentous breakthrough in 1954. The British and Australian groups constructed the first radio interferometers, which greatly improved the resolving power and therefore the ability to fix celestial co-ordinates of radio sources. Objects such as Cygnus A could be pinpointed on the sky to within a few minutes of arc. This enabled Graham Smith, working at Cambridge, to pin down Cygnus A to within one minute of arc. Using the then recently completed 5-metre reflector at Mount Palomar, Walter Baade and Rudolph Minkowski probed the sky near the radio sources. Cygnus A seemed to coincide with a disturbed 16-mag galaxy. Minkowski obtained a spectrum for the faint Cygnus A galaxy, quite an achievement at the time. Curiously, this spectrum was crossed by numerous bright emission lines, a sure sign of unusual activity. The lines showed a shift towards the red of $z = 0.057$, corresponding to a recession velocity of 17,000 km/sec. With the presently accepted value of Hubble's constant, and on the assumption that the Cygnus A galaxy obeys Hubble's Law, this gives a distance of 170 Mpc. The most astonishing facet of the discovery was the following: Cygnus A is the second brightest object in the radio sky; but its distance from our Galaxy is immense, and therefore its radio luminosity must be quite fantastic, 10^{38} watts to be precise. This is 10 million times higher than the normal background radio noise which can be picked up from a galaxy such as M31 or M33. Radio scientists were well and truly launched into extragalactic research with the discovery of such a titantic object that outshone nearer nebulae by many orders of magnitude.

Throughout the late 1950s and early 1960s radio astronomy was essentially in a cataloguing phase, analogous to the optical enterprises of Messier and the Herschels. In Britain, Australia and the United States radio telescopes methodically searched the sky, logging the position and brightness of each new-found source. Many of the brighter objects are still referred to by their numbers in the catalogues that first listed them. The *Third Cambridge* (3C) *Catalogue* first appeared in 1959, listing 471 sources; it was revised in 1962. Cygnus A is number 405 in this compilation, and accordingly is often referred to as 3C 405. As techniques improved so did the number of sources, so that the later *Parkes* (PKS), *Ohio* (O) and the *Fourth* and *Fifth Cambridge* (4C and 5C) catalogues list many thousands of sources.

The optical identification of radio sources is secured from examination of high-quality photographs, usually the Palomar Sky Survey. The area in the vicinity of the radio emission is scrutinized carefully, any plausible objects coincident with the radio position being noted. Except for the brightest objects, an accuracy of at least 10 seconds of arc in each co-ordinate is generally needed for worthwhile identification work. Interferometers now reach this resolution routinely, but this was not always so. Even by 1970 no more than 200 radio galaxies had been positively identified, but at the time of writing (1975) it exceeds 1000.

9.2 Optical properties of the radio galaxies

The strong radio galaxies, of which Cygnus A is frequently treated as a prototype, often show the unusual features which astronomers have come to associate with violent disturbances of one kind or another. Most radio-galaxy spectra show bright emission lines of highly excited gas such as ionized oxygen and nitrogen. In some cases the optical nucleus is either double or crossed by an opaque dust lane; this seems to be so for the galaxies associated with Centaurus A and Cygnus A. Other objects show peculiar trails and jets of gas; here we may cite M82, which matches radio source 3C 231, and M87, identified with 3C 274. In Cygnus A the optical spectrum displays signs of intense activity and energy. High excitation lines, such as [Fe X] and [O III], are intense and some thirty emission lines are identifiable on spectra taken by Baade and Schmidt. The nucleus of Cygnus A seems to be denser, hotter, larger and more energetic than any Seyfert galaxy nucleus. Great intensity in the emission lines aids the identification programme, since the majority of field galaxies do not show the strong lines in emission. There is also the question of determining a redshift, which is obviously easier if much of the energy is concentrated into a few lines. For example, 3C 123, with a redshift $z = 0.67$, is not only a radio galaxy, but it is also the most distant galaxy for which a redshift has been reliably determined.

In dealing with the classification of radio galaxy types it is necessary to extend the classical scheme put forward by Hubble. Radio galaxies fall into the following extra types:

(i) The D galaxies; these are elliptical with extensive surrounding envelopes.
(ii) Midway between types D and E is the DE galaxy.
(iii) Some radio galaxy nuclei, such as Cygnus A and Centaurus A (NGC 5128), are apparently double; these are called DB galaxies because of their approximate dumb-bell shape.
(iv) N–galaxies have brilliant starlike nuclei, although their outer structures are perhaps more easily discerned than in Seyfert galaxies.

Powerful radio galaxies are almost invariably variations on the giant ellipticals. In fact there are many cases where a radio source is identified with the brightest elliptical member of a cluster of galaxies. Since the normal ellipticals are generally almost devoid of gas, the fact that the radio ellipticals often show bright optical spectra, typical of gaseous nebulae, indicates that the presence of strong radio emission may be related to the excited gas in the nucleus. The widths of the emission lines certainly point to a surfeit of kinetic energy in the nucleus; in Cygnus A the velocities corresponding to the observed widths are about 400 km/sec.

The nuclei of N–galaxies vary in optical brightness, often violently. The galaxies identified with 3C 109, 309.3, 371 and the southern hemisphere source PKS 0521–36 have all shown sharp changes of brightness exceeding 1 mag. However, it is unlikely that any direct connection exists between these fluctuations in brightness and the extended double radio sources. In 3C 390.3 the components are now 100 kpc out from the nucleus, where they can no longer influence the nuclear intensity. The fact that variations have been recorded over periods of only a few years illustrates that the most luminous part of the nucleus in the galaxy concerned is only a few light years in diameter.

9.3 Properties of the radio emission

The radio emission from the radio galaxies has a very characteristic spectrum. Most sources have a spectrum in which the *flux density*, or received power, measured in (watts/m^2)/Hz, is proportional to $\upsilon^{-\alpha}$; here υ is the *frequency* at which flux density S is measured and the index α is a numerical quantity called the *spectral index*. For most extragalactic radio sources it is observed that

$$S(\upsilon) = k\upsilon^{-\alpha},$$

where k is a constant of proportionality. The units for S contain the bandwidth of the receiver, measured in Hertz (cycles per second, abbreviated Hz), because a 10-MHz bandwidth, for example, will pick up ten times as much power as a 1-MHz bandwidth at a given frequency, say 1420 MHz. Thus the received flux density is expressed as so many watts, per square metre of collecting area in the radio telescope, fore very Hertz of bandwidth, i.e. (watts/m^2)/Hz. The received powers are extremely tiny, so that a unit known as the *flux unit* or *jansky* has been introduced; in metric units it is 10^{-26} (watts/m^2)/Hz. Observed flux densities from 10 MHz to 10 GHz span a range 0.1–40,000 jansky. In order to measure a flux density of around 1 jansky a large collecting area—upwards of 100 m^2, and a large bandwidth—at least 1 MHz, would normally be necessary; even then the total received power will be 10^{-18} watts only, so that very-high-gain receivers are essential.

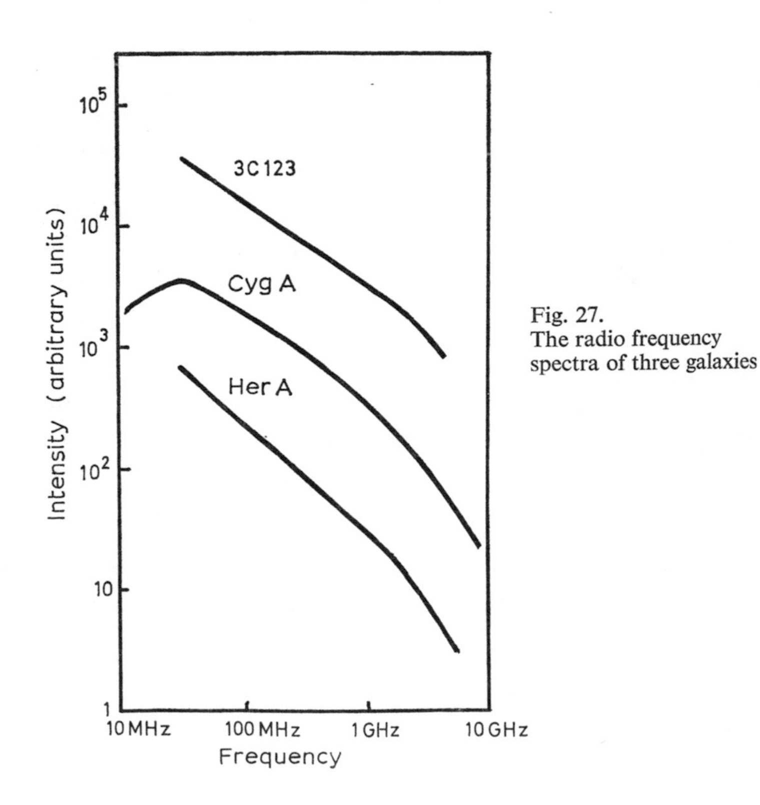

Fig. 27.
The radio frequency spectra of three galaxies

Our flux density law $S(\nu) = k\nu^{-\alpha}$ can be modified by taking logarithms

$$\log S(\nu) = -\alpha \log(\nu) + \log k.$$

This shows that a plot of the logarithm of received power $S(\nu)$ against the logarithm of frequency ν will be a straight line whose slope gives the spectral index α. Representative examples of radio galaxy spectra are plotted in this manner in Figure 27. It can be seen that in some cases the spectral index α is not constant over the entire spectrum; Cygnus A, for example, reaches a peak at about 20 MHz.

The spectral indices of most radio galaxies are in a range 0.6 to 0.9. About 60 per cent of all sources are bracketed by these values, a sign that we are probably dealing with the same emission mechanism in all of them. Once we have decided what this mechanism might be, it is possible to set about calculating the energy necessary to fuel a typical radio galaxy throughout its lifetime.

The most widely accepted hypothesis, and the one which most satisfactorily accounts for the observed radio spectra, is that synchrotron radiation is being generated. This type of electromagnetic radiation arises in the following way. When an electrically charged particle moves at speed across a magnetic field it experiences an electromagnetic force at right angles to the line of field and the direction of motion. This force causes the charged particle to make a spiral motion around the field lines. In turn the spiralling motion, which in reality is an acceleration whose direction is changing continuously, causes the particle to radiate electromagnetic energy across a broad range of frequencies. (All electrically charged particles emit energy when they are accelerated.) If the particle is travelling at close to the speed of light the energy is concentrated into a narrow beam that fans out into the instantaneous direction of motion (Figure 28). The energy released by this electromagnetic mechanism is called synchrotron radiation, because it was first recognized in laboratory particle accelerators known as synchrotrons.

In astrophysical situations the charged particles are electrons and the celestial magnetic field strengths are 10^{-10} to 10^{-7} webers per square metre; the latter figure is about one-thousandth of Earth's magnetic field. There are vast populations of electrons with a wide range of energies, so that at any one time we are observing the combined effect of all of them. The highest energy electrons radiate at the highest frequencies and lose energy the quickest. Therefore, as a

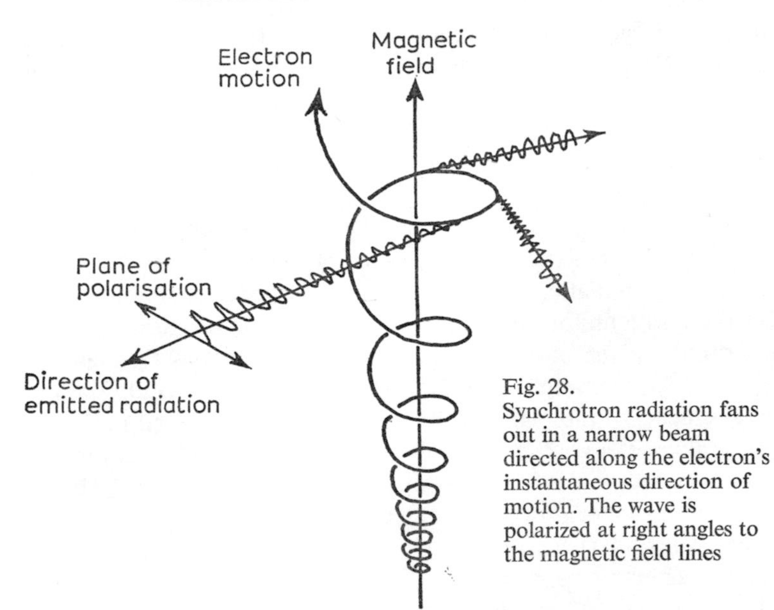

Fig. 28.
Synchrotron radiation fans out in a narrow beam directed along the electron's instantaneous direction of motion. The wave is polarized at right angles to the magnetic field lines

radio source ages, it will gradually get depleted of the most energetic particles, and the flux density at high frequencies will decrease. This may explain the curvature in the spectrum of objects like Cygnus A.

To produce the straight-line spectrum with index α requires an underlying population of electrons which also follows a power law

$$N(E) = kE^{-(2\alpha+1)},$$

where N(E) is the number of electrons with energy E. We see from this that the characteristic spectra of radio galaxies reflect an underlying order in the energy spectra of the electrons. It is a major task of any theory of extragalactic radio sources to account for the relatively simple law which expresses the way in which the available energy is distributed among the high-speed electrons.

Further evidence favouring the synchrotron radiation interpretation comes from the study of polarization in the radio waves. The usual situation in radio astronomy is that a small percentage (up to 10 per cent typically) of the radiation at a given wavelength is found to be vibrating in one plane, or in other words, the radio waves are partially polarized. The synchrotron emission process results in the emission being linearly polarized in a direction that is perpendicular to the lines of magnetic field. It might, therefore, be expected that if there are large-scale and well-aligned magnetic fields in the extragalactic radio sources, some degree of linear polarization should be present in the received waves. This is in fact the case. In 1962 it was found that the radio waves from Cygnus A were partially polarized. Subsequent studies from many hundreds of sources have shown that most of them are partially polarized.

The degree of polarization and the angle at which maximum polarization is observed in the received wave varies with the wavelength. When radiation traverses a region of space containing both electrons and a uniform magnetic field, the wave becomes disturbed such that its plane of polarization rotates as it travels through the medium. This rotation is called the Faraday effect. By taking measurements of the plane of polarization in the received radiation at several wavelengths, it is possible to infer how much rotation has been impressed on the radiation since it left the radio source. This in turn provides a means of probing the average density of free electrons in the region of space between the radio telescope and the radio source. Furthermore, the polarization data can be used to determine the structure of the underlying magnetic field, as shown in Plate 18.

9.4 Radio galaxy architecture

The structures of the radio components and their relation to the optically visible galaxies are investigated with high-resolution radio telescopes. Since the necessary resolving powers for making full maps of the radio structure of many sources were not available until 1966 (the date of completion of the One Mile Telescope at Cambridge), progress lagged behind the cataloguing and spectrum research. However, this situation no longer holds now that synthesis telescopes are at work in Britain, the Netherlands, Australia and the U.S.A. Most of the radio sources are less than a minute of arc in size, but modern interferometers are able to see individual details as small as 1 second of arc in diameter; by linking telescopes on different continents to form one very-long-baseline interferometer it is even possible to discover structures less than 0.001 seconds across, although these cannot be mapped in detail.

The usual method of displaying the structural information is to use a contour map. Individual contours enclose regions within which the radio emission exceeds a certain flux density. (Just as a contour line on a geographical map encloses regions within which height

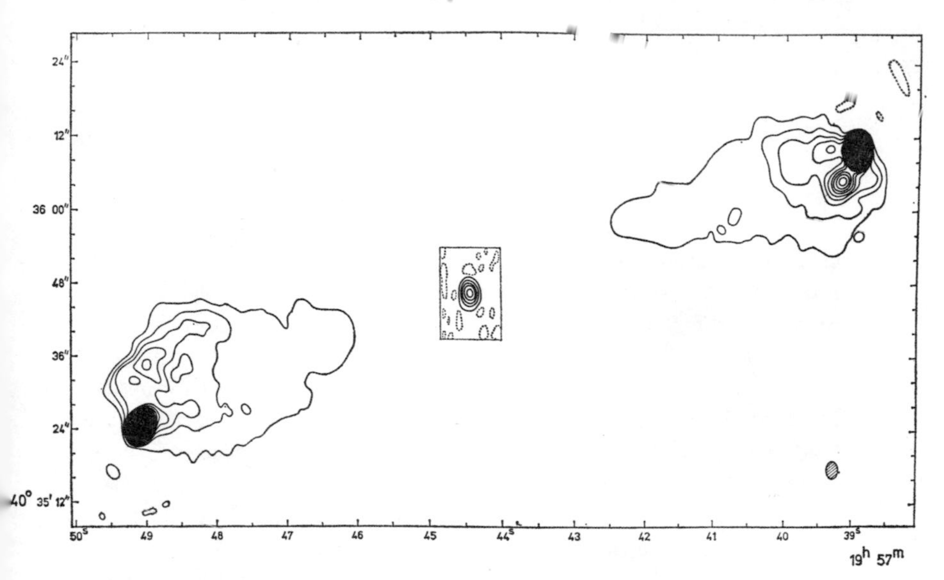

Fig. 29. The radio galaxy Cygnus A mapped at 5 GHz by the Cambridge 5-km telescope. Faint radio emission has been detected from the optical galaxy. In the main components are two compact sources which are here shown by solid ellipses

exceeds a specified value.) The radio contour maps show that most radio galaxies consist of two large clouds of radio emission symmetrically disposed on either side of the main galaxy. So the radio emission originates in invisible regions that are generally far beyond the confines of the associated galaxy (Figure 29). An impressive feature of the radio clouds is their immense size. They may be found 10–100 kpc from the galaxy, and they frequently measure tens of kpc along their main axis. This, of course, means that the plasma clouds may well be much bigger in volume than is the galaxy with which they are associated. In a few extreme cases they are as colossal as a large cluster of galaxies.

The classical double structure of radio galaxies strongly suggests that they are caused by galactic explosions. During these events the fast electrons and the magnetic field responsible for the synchrotron radiation are thrown out in opposite directions. Radio structure maps have shown that the edge of the plasma cloud furthest from the optical galaxy is usually the brightest. This is due to compact condensations at the peripheries of the radio clouds, and there may be a shock wave caused by an interaction between the clouds and the matter in extragalactic space.

When the structural information is combined with the synchrotron radiation hypothesis we can make an estimate of the relativistic energy that must be stored in the plasma clouds in order to yield the observed radio luminosities. By relativistic energy we mean the energy needed to produce the magnetic field and to accelerate the electrons to almost the speed of light. It turns out the energy requirement is a minimum (i.e. the cosmic power station is working at maximum efficiency) if half the energy is put into creating the magnetic field and half into accelerating fast particles. We do not know how Nature has arranged matters in the radio clouds, but by assuming that the available energy is split equally between particles and field we get an idea of the absolute minimum amount of energy stored in a typical radio source. As we might have anticipated from the fact that the luminosities are very large—up to about 10^{38} watts—the requisite store of energy is immense. In Cygnus A it probably amounts to a minimum of 10^{52} joules, and for most sources it exceeds 10^{50} joules. An appreciation of the magnitude of these figures may be gained by stating that the total rest-mass energy of our sun, that is the energy that would be released by utterly annihilating all its matter, is 2×10^{47} joules. We shall return again to the problems raised by the existence of such large energies. For the present, note that these huge energies are one of the most intriguing properties of the extragalactic radio sources.

9.5 Some well-studied radio galaxies

Cygnus A

Cygnus A is widely regarded as the prototype of all radio galaxies, because it is among the most powerful and energetic known, and, lying at 170 Mpc from the Galaxy, it is one of the nearest available for a detailed study. Unfortunately it is in the plane of the Milky Way and this has hindered optical studies. However, it is known to have a highly excited nucleus and it is the brightest member of a rich cluster of galaxies. As a radio source it is intermediate in physical size, stretching for nearly 100 kpc between the extremities of its two components. Sir Martin Ryle and Phillip Hargrave published maps of the radio structure in 1974, made with the Cambridge Five-Kilometre telescope. These show (one is reproduced as Figure 29) that each component is composed of several condensations of high brightness, with a compact hot-spot at the front end of each component. They have also investigated the magnetic field structure by mapping the polarized radio emission. The magnetic field breaks into several tangled regions, although an overall field running parallel to the major axis of the radio source can be discerned. Perhaps the most significant aspect of these magnificent maps is the discovery of the numerous compact blobs, in which a great deal of energy is concentrated.

Centaurus A

The southern hemisphere source, Centaurus A, is the nearest of the radio galaxies, being only 4 Mpc away. It is identified with galaxy NGC 5128, which is crossed by a conspicuous central dust band. Centaurus A is among the intrinsically largest radio sources and its radio contours span 10° of sky, so that its structure was investigated before the high-resolution telescopes had come into action. The extended radio emission stretches through 600 kpc of intergalactic space. Despite this large size it is 1000 times less powerful than Cygnus A. Polarization measurements show that much of this region is permeated by a smooth magnetic field, having a few zones which exhibit small-scale structure. Besides the two giant radio clouds there is a second, much smaller pair of clouds which have not yet emerged from the outer boundary of the galaxy. This central radio source contributes about 20 per cent of the total emission. Optical research has shown that the axis of rotation of the galaxy is perpendicular to the dust band round the galaxy; the radio components almost certainly emerged from this galaxy along the rotation axis.

Virgo A (M87)

The galaxy with a jet, M87, is also a strong radio source, but of a different type to the classical doubles. It consists of an intense central radio source which coincides with the galactic nucleus with *two* jets of radio emission darting out on each side. One of these coincides with the stream of bright knots visible on photographs. Both the optical and radio emission from these condensations is probably caused by synchrotron mechanism.

3C 33

This is an intriguing double radio source, identified with an elliptical galaxy. According to maps made in California and Cambridge it has two small components which are separated by 100 kpc—sixty times their own diameter. This implies that the expansion speed of the plasma clouds on release from the galaxy must have been much smaller than the velocity of light, because the cloud expansion velocity must be far lower than the velocity at which they are separating. This, then, is a fine example of a tightly contained radio source. The puzzle is deciding what helps to hold it together so well.

3C 295

Mention should be made of the most distant radio galaxy for which there is a reliable redshift. Minkowski showed that the galaxy identified with the double source 3C 295 with a redshift $z = 0.461$, which would imply a distance of 1300 Mpc. It has the highest radio luminosity known.

3C 465

Not all radio sources are as well organized as the ones discussed above. The inner double structure of Centaurus A suggests that energy release may occur more than once. This theme is strengthened in the case of 3C 465, a complex of radio components spreading in an arc through a cluster of galaxies (Figure 30). Here more than one galaxy may be implicated, or the morphology could reflect repeated events while the galaxies swarmed around each other in the cluster.

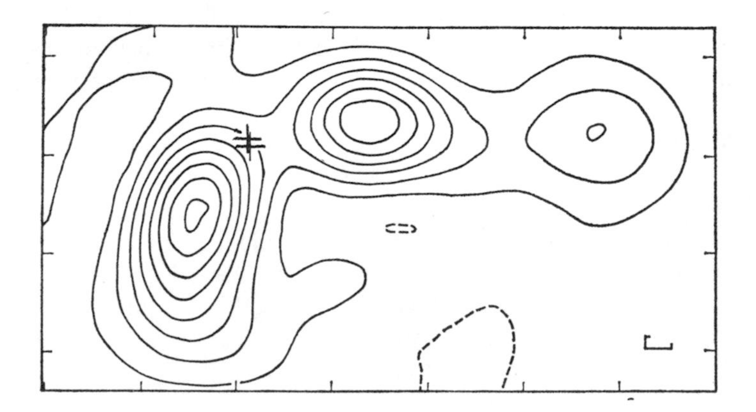

Fig. 30. The enormous radio source 3C 465 spans a cluster of galaxies centred on the pair of crosses

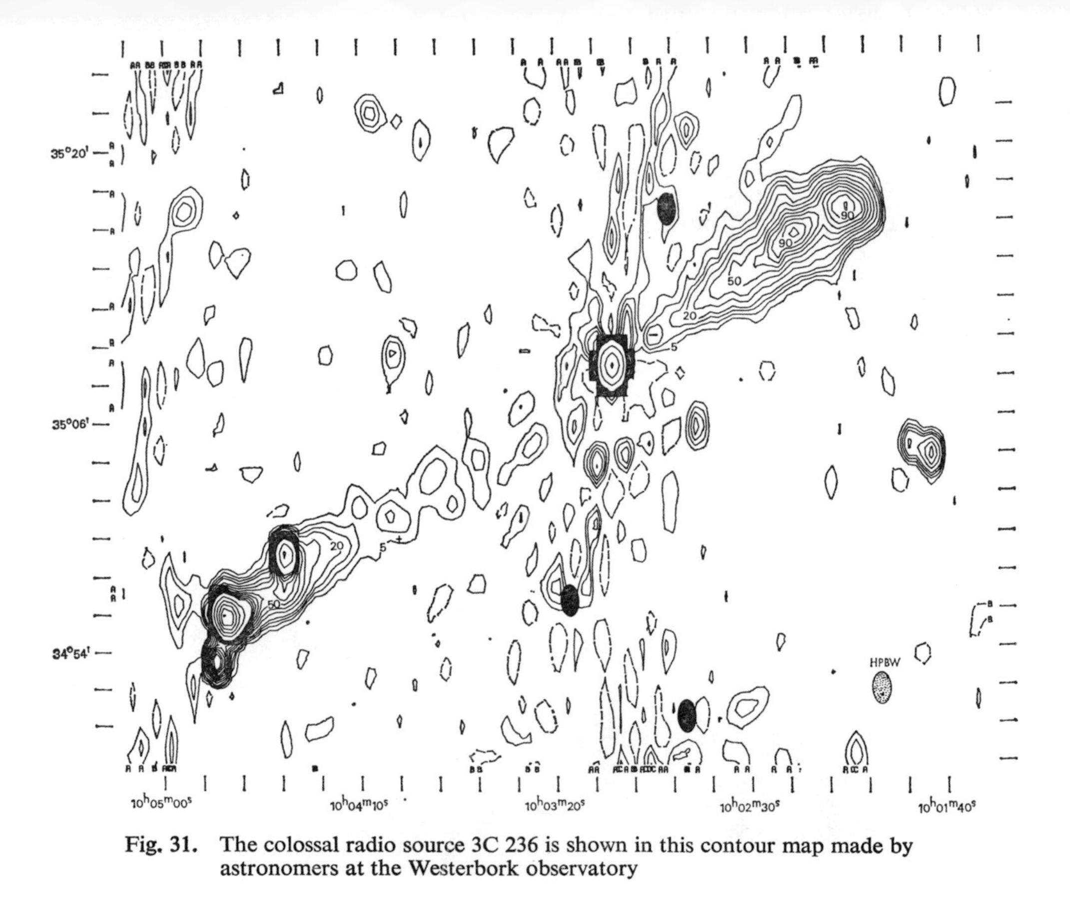

Fig. 31. The colossal radio source 3C 236 is shown in this contour map made by astronomers at the Westerbork observatory

3C 236

This object is remarkable because it is the largest radio galaxy known. In 1974 Harry van der Laan and his associates at the Leiden Observatory announced that the radio structure of 3C 236 covered 0.5° of sky. They made this discovery with the Westerbork Synthesis Radio Telescope, which had then recently been fitted with highly sensitive receivers to work at 50-cm wavelength. The instrumental configuration employed was just about the best available for plotting very faint distributions of surface brightness. As a result they found that 3C 236 was much more extensive than anyone had previously supposed. It displays the classical double structure, possessing as it does two trumpet-shaped radio components emanating from a faint galaxy (Figure 31). From the galactic redshift and the observed angular size, a physical extent of 5.7 Mpc has been inferred for this source. This is over fifty times as big as Cygnus A, and is about ten times larger than anything else that had been discovered up to 1974.

Another giant, DA 240 (Plate 20), which extends 2 Mpc end to end was found in the same search for large galaxies. Both sources have energy requirements that are among the largest known. They present profound problems to theorists interested in the evolution of radio sources, because the vast size implies an age of tens of millions of years at least. It is difficult to conceive of processes that cause galactic nuclei to hurl balls of relativistic plasma across a distance over ten times as far as the Andromeda nebula. 3C 236 and DA 240 are thought to be the largest known objects in the universe.

9.6 Radio waves from tadpole galaxies

In 1968 Martin Ryle and Michael Windram published maps showing the radio emission from the galaxies in the Perseus cluster. The Seyfert galaxy NGC 1275 is identified with a compact radio source 3C 84. Not only is NGC 1275 the optically brightest galaxy in the Perseus cluster, but its associated radio source 3C 84 is also the brightest in the vicinity. The radio source is only a few parsec in diameter. The whole Perseus cluster is enclosed by a vast halo of low-frequency radio emission, extending hundreds of kiloparsecs (Figure 32). Ryle and Windram also determined the structure of radio sources associated with NGC 1265 and IC 310, two further galaxies in the cluster. They suggested that high-velocity gas ex-

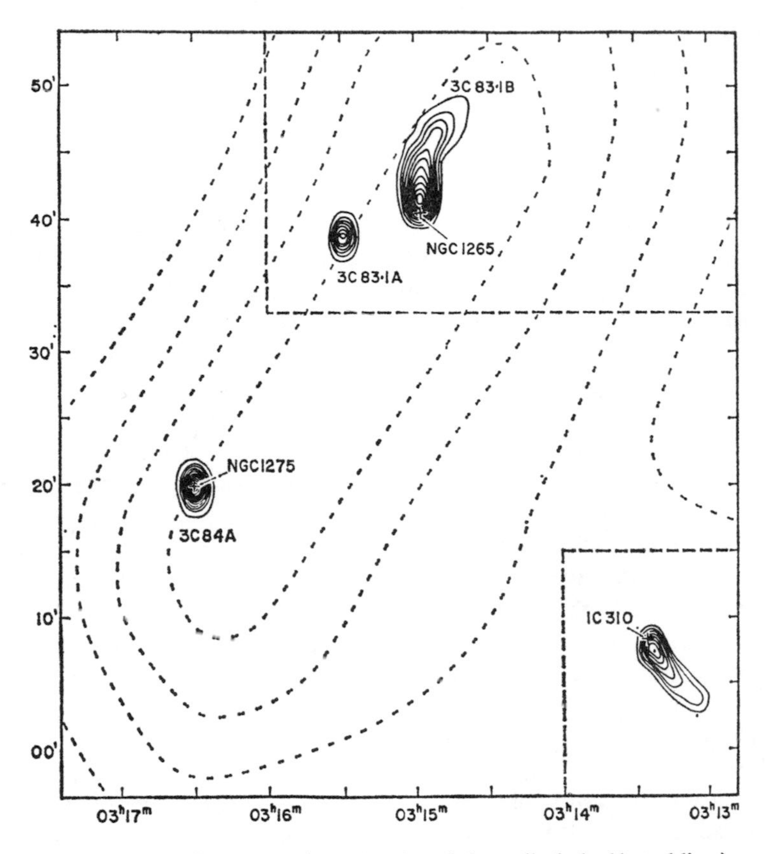

Fig. 32. Radio sources (solid lines) and the radio halo (dotted lines) in the Perseus cluster of galaxies

pelled from NGC 1275 had crashed into two neighbouring galaxies and excited radio emission within them. Supporting this was the observation that the sources near to NGC 1275 had bright 'heads' coincident with their galaxies and 'tails' pointing away from 3C 84; just the type of structure that would be produced by a wind of fast particles rushing past a galaxy (Plate 18).

Further work on the Perseus cluster by the astronomers at Leiden Observatory led to the discovery of another head-tail source in the vicinity, but this one did not line up with 3C 84. Investigation led them to the conclusion that none of the sources near NGC 1275 is directly triggered by 3C 84. Instead Harry van der Laan and his associates put forward the view that the head-tail or 'tadpole' galaxies are yet another type of radio source, associated with clusters of galaxies. They found some further examples, such as 3C 129/129.1 later, while a research student at Cambridge, Martin Willson, drew attention to another candidate, 5C 4.81 in the Coma cluster of galaxies.

The half-dozen tadpole galaxies are all in clusters. They are characterized by a compact source very close to a galaxy and a long sausage-shaped extension that juts out into intergalactic space. High-resolution maps from the Dutch synthesis telescope at Westerbork show that the emitting plasma emerges as two streams from the galaxy; these two streams are swept back and merge to form the long tail. A suggestion put forward as an explanation of the tadpole galaxies is that they are moving rapidly through the hot gas which permeates galaxy clusters. As the cluster gas brushes past the galaxy it gathers together and propels along two emerging streams of plasma to form the tail of the plasma. Presumably, an ordinary double source would have formed in the absence of the supposed motion. Curiously enough, the Perseus cluster, which contains three tadpoles, has galaxies moving at up to 3000 km/sec, relative to the cluster centre, according to Chincarinni and Rood.

9.7 Problems for the model makers

Even though radio galaxies have been known since 1954, the theoretical approach to the problems has not progressed beyond the model-making stage. In a nutshell, the basic properties of radio galaxies are: intense power; great energy; generally double structure; relatively compact components; components have travelled large distances (100 kpc) from their parents; and strong evidence that the galactic nuclei are highly disturbed. To make any progress whatsoever the problem must be treated in a number of stages, and we can summarize these in the following paragraphs.

(1) *The energy problem.* Available evidence on activity in the nuclei of galaxies suggests that the energy needed to make a radio source is generated there in the first instance. It is necessary to establish from where this energy comes, and also to find out whether the energy is released directly as energetic particles and magnetic field, or whether some intermediate stages are involved in the production

of the plasma clouds. The energy problem is central to much of extragalactic astrophysics, and we shall return to it in Chapter 12.

(2) *The morphology.* Most radio galaxies are doubles or distorted doubles. It is difficult to understand why the clouds should be ejected as two roughly equal pairs with opposing velocities. Dynamical effects, or the influence of magnetic field, could be responsible for the bifurcation. Observations of compact radio sources made with very-long-baseline interferometers show that in some cases the doubling has already taken place while the nascent radio source is still inside a compact galactic nucleus.

(3) *Keeping sources alive.* Once the energy has been released and a pair of plasma clouds created, many factors conspire to kill off the radio emission. One is expansion: radio sources like Cygnus A must have grown in size from something smaller than a galactic nucleus to a cloud that is bigger than a galaxy. If a cloud of relativistic matter is simply thrown into the near-vacuum of extragalactic space, the resulting expansion causes a great loss of energy. Instead of radiating, the particles use up their energy in accelerating the outer boundary of the cloud. There must be some way by which radio clouds conserve their relativistic energy, or at least replenish it. Extra energy could be fed into the component long after its creation by means of low-frequency radiation from the galactic nucleus. Or the components might have a population of energetic, invisible, collapsed objects such as neutron stars, pulsars or black holes, inside itself; these exotic creatures could accelerate relativistic particles for the duration of the radio galaxy's lifetime. Yet another idea is that the components are seeded with a vast amount of cold gas. This carries large amounts of energy, by virtue of its momentum, as it speeds through extragalactic space, and possibly this energy of motion can be turned into higher energies through turbulence in the plasma. These are just a few of the many possibilities that have yet to be explored rigorously. All we can say is that there are plenty of ways of getting more supplies of energy, but we do not know which of them are correct.

(4) *The containment problem.* The radio component consists of electrons whirring round the magnetic field lines at close to the speed of light. Left to itself, this relativistic plasma, as it is called, would

expand into the vacuum of intergalactic space at about one-half the speed of light. The existence of tightly contained radio sources, such as 3C 33, shows that something acts to restrain the expansion of the radio components. There are several ways of tying the components together: gravity within the component, perhaps due to the presence of a few very massive objects, could do this; or there may be an external pressure caused by heat or magnetic effects which squeezes the plasma bubble; another possibility is that the clouds are weighed down with vast amounts of cold matter—this would expand rapidly and would restrain the relativistic matter from doing so as well; yet another idea is that the passage of the radio component through the intergalactic medium creates a shock wave which then acts as an envelope for keeping the component together. Now that X-ray emission has been discovered from rich clusters of galaxies and interpreted as emission from a hot cluster gas, the possibility that external thermal pressure or shock waves hold the radio sources together is increased.

(5) *The distribution of particle energy.* Even if a means for creating energy is thought through, there are problems associated with feeding this highly disorganized energy, which may come from a dramatic explosion, into the neat particle-energy spectrum, which is a simple power law. It would be somewhat easier to invent models of particle acceleration if these spectra were curved.

It can be seen from the above summary of radio galaxy problems that a bewildering array of choices faces the model makers. For a time some particular scenario will enjoy popularity, only to be overtaken when someone comes up with a new idea for getting further with some of the other choices. It is probable that such complex problems will only yield after years of hard work, rather than a flash of insight in a bathtub, that traditional source of great scientific breakthroughs! In the meantime the world's synthesis telescopes are producing a steady stream of maps and these can be used to strengthen or destroy the theoretical research, depending on one's inclination. There is the possibility that a range of radio source types exists, and any given source is the end-product of several different mechanisms working together or in competition. If we had tried to explain all the stars in the sky in terms of one particular model of stellar structure we should not have succeeded. On the other hand, we do not yet have enough information on radio galaxies to hand to see a clear pattern of different species emerging.

10 · Quasi-stellar objects

10.1 The discovery of quasars

The remarkable properties of strong radio sources such as Cygnus A added impetus to general programmes aimed at identifying and mapping as many radio sources as possible. Interferometers operated by astronomers at the Jodrell Bank station of the University of Manchester had shown that some of the brightest objects listed in the 3C catalogue had angular diameters of 1 arc second or less. Around 1960 observers considered it especially remarkable that a few radio sources such as 3C 48, 286, 196 and 147 defied all efforts to resolve their structures. The assistance of optical astronomers was enlisted and so up the mountain to the great 5-metre reflector of the Hale Observatories went Allan Sandage. He wished to see what objects were visible in the fields of the first three sources listed above.

T. A. Matthews pored over Sandage's photographs and found in each case that a starlike object appeared in the error rectangle defined by the radio astronomical position measurements. Were these the long-sought 'radio stars', members of our own Galaxy? Further colour measurements and spectra were needed to clarify this question. By October 1960 Sandage had this date for 3C 48. The star apparently associated with 3C 48 certainly had strange properties. Its colour resembled a white dwarf or old nova star; also it showed brightness fluctuations in a time scale of about one day. So, it was argued, the object could not be much more than one light-day across. An interesting galactic star had been found and it had a weird spectrum which defied explanation. This 'official' view held sway until 1963, a tumultuous year during which the entire course of extragalactic astronomy and physical cosmology changed abruptly.

In late 1962 Cyril Hazard, Mackey and Shimmins were using the 65-metre radio telescope at Parkes, New South Wales, on a project to observe the occultation of radio sources by the moon. The purpose of this lunar occultation technique, developed for accurate positional work by Hazard, was as follows: the position of the edge (or limb) of the moon in the sky at any time is known with very great precision. Consequently, when the moon slides across a more distant object, accurate measurements of the time at which the lunar limb chops out the radiation from the object and the time at which the signal subsequently reappears provide a very accurate position for the radio source. Apart from finding positions there is further information which was the original *raison d'être* of the radio occultation programme. The occulted object does not disappear and reappear instantaneously. Rather, the limb of the moon acts as a diffraction screen and disturbs the electromagnetic waves from the radio source, so that a fluctuating signal, known as a diffraction pattern, is received by the radio telescope. From the observed fluctuations one can infer something about the structure and size of the radio source. So the lunar occultation technique is a cheap and effective way of finding positions and structures with a resolution of 1 arc second or better. The big disadvantage, of course, is that you have to wait until the moon obligingly swims across the sources of interest; this is absolutely limited to declinations between $-30°$ and $+30°$.

In late 1962 the moon happened to occult 3C 273 three times, which enabled Hazard to pinpoint it to within 1 arc second. Furthermore, he and his colleagues found that it possessed a double radio structure, and that one radio component coincided with a 13-mag star.

So reliable was the identification of 3C 273 with a star thought to be in fact, that Maarten Schmidt obtained a spectrum, and he also discovered a faint bluish jet extending out from the star along the main axis of the radio source. Superficially, the spectrum resembled that of 3C 48: several broad emission lines that could not be identified with atomic transitions recorded in the laboratory. What kind of stars were these?

The solution, when it came, was of far-reaching consequence. Maarten Schmidt showed that he could interpret the spectral lines in terms of a substantial redshift. Four of the emission lines slotted into the famous Balmer series of hydrogen on adoption of 0.158 as the redshift for the light from 3C 273; then the remaining lines could be correlated with known transitions. Jesse Greenstein and T. A.

Matthews cracked the mystery of 3C 48 on learning of Schmidt's work. All its lines fell neatly into place with an assumed redshift of 0.367.

The fun was only then beginning. The end is not in sight even today. For a start, the redshifts of 3C 48 and 273 implied distances of 1100 and 480 Mpc respectively, if the Hubble Law of redshift-distance was to be believed. But 3C 273 is of 13 magnitude—bright enough to be visible by eye in a telescope. Its absolute brightness must be over 100 times that of the brightest known galaxy for the apparent magnitude to be as bright as 13 mag after a 480 Mpc journey. Both 3C 48 and 273 are prototypes of a class of extragalactic sources known as *quasi-stellar objects* or quasars. The personality trends of the family are high redshifts and a starlike appearance optically. Many of them are also strong radio sources.

In the years following 1963 progress was rapid. Eight more were identified in 1964, including 3C 147, which had $z = 0.545$, then a record redshift. Ten years later the number of known quasars stood at many hundreds. Maarten Schmidt gradually worked up to larger and larger redshifts, as experience in identifying the spectral lines in high-redshift objects increased. A crowning achievement at the time was the fantastic shift of $z = 2.012$ found for 3C 9. In this object a fundamental transition in hydrogen, the Lyman-alpha line, is red-shifted from the ultraviolet at 1216 Å down to 3660 Å in the visible. This was the first recording of Lyman-alpha in a remote celestial body, and the result had important implications for the properties of the intergalactic medium.

With the large redshifts we can no longer write the classical relation between redshift z and velocity v:

$$z = v/c.$$

This would admit velocities exceeding that of light, c. The special theory of relativity shows that light is shifted from λ to $(1 + z)\lambda$ at velocity v in accordance with

$$1 + z = \left(\frac{c + v}{c - v}\right)^{\frac{1}{2}},$$

a relation illustrated in Figure 33. Once z exceeds about 0.5 the simple $z = v/c$ approximation is misleading. For 3C 9, $z = 2.012$, and the corresponding recession speed is about 80 per cent that of light.

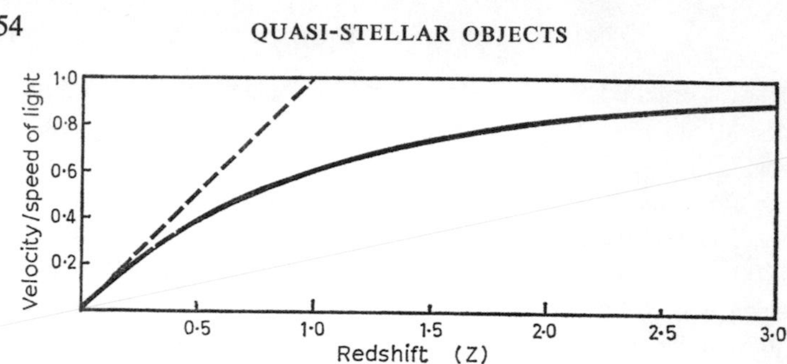

Fig. 33. The variation of redshift with velocity is shown as a solid line. The simple relation is shown as a dotted line, which deviates markedly from the true behaviour when z exceeds about 0.5

Some of the brighter quasars can be traced on photographic patrol surveys going back to the beginning of the twentieth century. This old material frequently shows that many quasars have fluctuated in brightness over the years, even by two or three magnitudes. More recent studies of the fainter objects have disclosed that significant variations in a matter of days or weeks are not rare. The luminosity of the quasar as a whole varies, and the diameter of the main zone of energy release can only be a fraction of a parsec according to the light variations. If we bear in mind that this region could be pouring out to the cold universe 100 times as much as a giant galaxy, we see an enormous barrier to knowledge looming up. How do the laws of physics permit a quasar to pack such a mighty punch into so small a volume?

Another important discovery made by optical astronomers was that some of the high-redshift quasars have numerous absorption lines in their spectra. Perhaps the most puzzling aspect of the absorption-line patterns is that some spectra can only be explained by the hypothesis that several redshifts are present in one and the same spectrum. In certain cases it has been suggested that as many as half a dozen distinct systems of absorption lines exist simultaneously. The quasar PHL 957 has a rich absorption-line spectrum. Its emission lines yield a single redshift of 2.69. Image-tube spectra have revealed over eighty absorption lines, and according to John Bahcall and Paul Joss (Institute for Advanced Study, Princeton) five absorption-line redshifts are present. They are $z = 2.67$, 2.55, 2.54, 2.31 and 2.23. Another object with a complex spectrum is 4C 05.34, which has an emission-line redshift $z = 2.88$ and well over

100 detectable absorption lines with at least six distinct redshifts varying from z = 2.87 down to 1.78. The absorption lines are often very narrow, implying a spread in velocities of only a few tens of km/sec in the gas causing the emission.

The most straightforward interpretation of the multiple line systems is that the quasar has ejected several successive shells of absorbing material. The velocity of this material relative to the quasar affects the redshift of the sets of absorption lines. In some cases, where there is a large difference between emission and absorption redshifts, it is necessary to postulate that the absorbing shell is rushing away from the quasar at close to the speed of light. A different possibility is that the light from the quasar has had to traverse several extremely distant galaxies during its long journey. Each time it passes through a galaxy, the interstellar matter within it impresses a set of absorption lines at the redshift of the galaxy. The observed absorption redshifts would place the intervening galaxies far beyond the reach of optical telescopes. At present it is not possible to distinguish these two hypotheses with certainty. However, it is possible that the richest absorption spectra contain lines of Lyman-α at 50 or even 100 different redshifts. If this interpretation is correct it favours the idea of absorbing clouds near to the quasar.

10.2 Observations of radio emission from quasars

Quasars were, of course, discovered during investigations of their radio emission. Many of them are indistinguishable from radio galaxies on the basis of their radio properties alone: they show the classical double structure, synchrotron spectrum and some polarization. Examples are 3C 9 and 3C 47 (Figure 34). But quasars can be highly compact radio sources and this discovery put much impetus into the development of interferometers. Longer and longer baselines were tried until eventually intercontinental separations were in use. Many quasars have a radio source coincident with the optical object which is less than 0.1 arc seconds in size, and some possess individual components smaller than 0.001 arc seconds.

In 1969 a link-up between a 64-metre telescope in California and a 26-metre telescope in Canberra, Australia, yielded a baseline of 81 million wavelengths at the observing wavelength of 13 cm. This separation is sufficient to resolve clearly structures exceeding 0.001 arc seconds in size. At each telescope the data are recorded onto

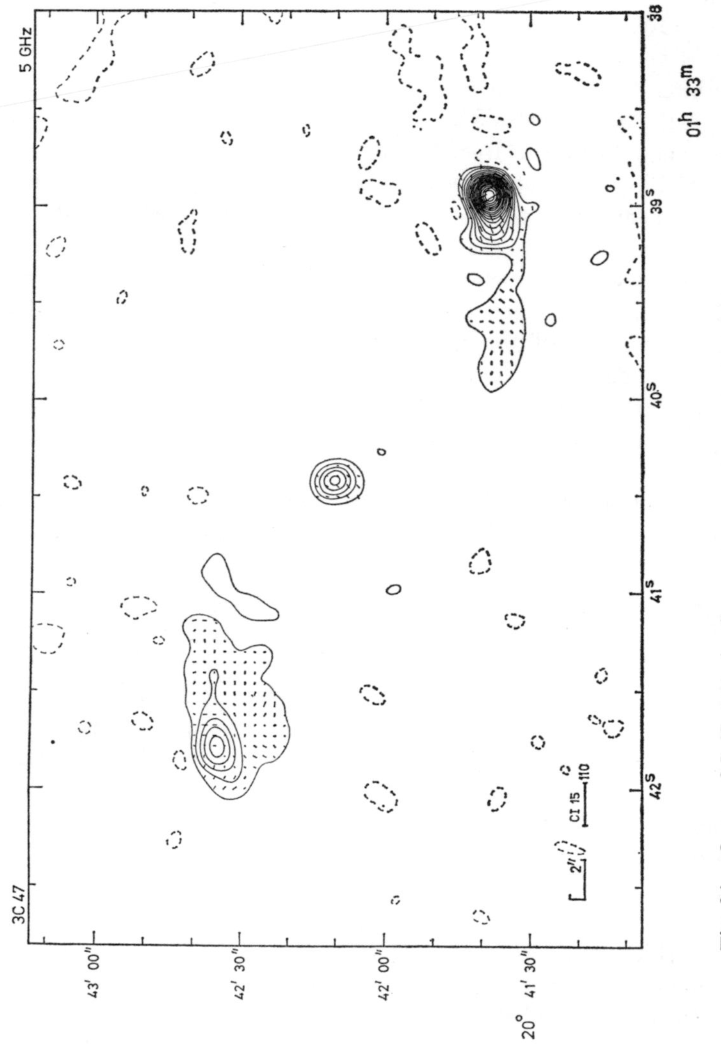

Fig. 34. Quasar 3C 47. Polarization vectors are shown as short bars

magnetic tapes along with accurate time marks from an atomic clock. Correlation and analysis of the data are performed by bringing the tapes to the same location and analysing them in a digital computer. The clocks at the two observatories can be synchronized by means of radar pulses bounced off the moon.

Observations with the Canberra-California interferometer showed over fifty sources, mostly quasars, with substantial structure below 0.001 arc seconds in size. A few of these were not resolvable at all even with the huge baselines available. Subsequent observations at shorter baselines (25 million wavelengths) have filled in the picture to a limited extent. Broderick, Kellerman, Shaffer and Jauncey claim that the long baseline information shows that quasars such as 3C 273 and 279 have complicated compact structures, and this is typical of many of the small diameter sources. Furthermore, there is much evidence, particularly for 3C 279, that the structure changes on a time scale of only a few weeks. This effect could be due to individual knots of emission moving apart rapidly, or, more probably, to one or more small components changing its radio luminosity. At one time it was even thought that 3C 279 contained components moving apart at more than the speed of light; however, a correct relativistic analysis of the data shows that this is not the case, although the velocities implied are extremely high.

An early discovery in the investigation of quasi-stellar radio sources was of the radio outbursts exhibited by the compact objects. In 1965 W. A. Dent of the University of Michigan found that the compact component of 3C 273 had increased its flux density at high frequencies by 40 per cent in three years. Subsequently, several examples of increased radio brightness were noted. Generally the intensity rises to a maximum, which may be several times brighter than average, and then it decays again. The maximum intensity occurs at successively later dates as the observing frequency is decreased, so that any new variable source is usually first detected at high frequencies. A common theme in the variable sources is for a large outburst to last a few months and then to be followed by another burst within a year or so. Among the most active quasars in this respect we may list 3C 273, 279 and 454.3. Similar behaviour occurs in the Seyfert and N–galaxy nuclei sources 3C 84, 120 and 371. The radio variables also show sharp changes in their optical magnitudes, although fluctuations at optical and radio frequencies do not appear to be correlated.

The feverish activity at radio frequencies can be explained by clouds of high-energy electrons being periodically ejected from the nucleus of the quasar. As the cloud of electrons expands it becomes transparent to radio waves at progressively lower frequencies. At the start of a burst only high-frequency radiation can pierce the electron cloud, so that a maximum intensity is registered at high frequencies. The position of this maximum drifts to lower frequencies as the electron density increases and the plasma becomes less opaque. The energy source which releases the clouds of electrons and magnetic field remains mysterious. Its size cannot be very great because the variations take of the order of one year. The compact radio sources within quasars are a variation on the theme of unusual activity in the nuclei of galaxies.

10.3 Mysteries of quasar redshifts

The abnormally large redshifts of quasars are their most intriguing feature. Other problems traditionally associated with the quasars stem principally from this property. Immediately the large redshifts were discovered the question as to whether the quasars obey the law of galactic redshifts (the Hubble Law) began to cause much heart-searching. Quasars with redshifts of 2.0 or so soon became commonplace, implying recession speeds of four-fifths the velocity of light. Once velocities as great as this are attributed to expansion in accordance with Hubble's law, enormous distances—of order 10,000 Mpc—result. The very fact that we can see the objects with optical telescopes and detect their radio emission implies colossal optical and radio luminosities if the distances derived from redshifts are accepted. Not surprisingly there has always been a number of astronomers who questioned an argument that resulted in intractable energy problems.

When redshifts for about 200 quasars became available, G. R. and E. M. Burbidge made a statistical analysis of the distribution of redshifts. Their histogram (a plot of the number of quasars with redshift in a given range against redshift) appeared to show that the redshifts cluster around certain values. They seemed to be quantized in units of 0.06, with a strong clustering around $z = 1.95$. Furthermore, it appeared that every quasar with absorption lines had one set at $z = 1.95$. These discoveries of quantization and clustering in the values caused a great deal of excitement because such behaviour implies preferred values and is not consistent with a simple law relating redshift and distance. Also, it has implications for the evolution of quasars; the large number at z = 1.95 could indicate an epoch when quasar formation was especially likely. It is now known that these results are principally due to selection effects and the use of small samples for the statistical analysis. Subtle selection effects mean that quasars are easier to detect and their spectra easier to analyse in certain redshift bands. The result on the absorption lines at z = 1.95 no longer holds. An analysis, in 1972, of nearly 400 redshifts, by G. R. Burbidge and S. L. O'Dell, failed to confirm the preference among the redshifts for multiples of 0.06. The addition of more redshifts has tended to smear out the clumping earlier noted at z = 1.95. So one set of mysterious properties at least has been laid to rest.

Until about 1973 it was thought that a strange 'cut-off' existed in quasar redshifts. Only one had been found with a redshift greater than 2.5, which led to suggestions that something, perhaps intergalactic hydrogen, was veiling their light. Another suggestion was that no quasars had formed in remote regions of the universe where z exceeded 2.5. However, quasars with redshifts beyond 2.5 were found eventually: 2.88, then 3.4 and then 3.53 being discovered in turn. The highest redshift objects are neutral or reddish in colour. Since quasar searchers had generally gone for the intense blue star-like images, high redshift quasars may have been discriminated against if they are generally red. This is yet another example of the care needed in interpreting observational data; the apparent cut-off may simply have been due to the method by which candidates were selected for quasar searches.

10.4 Quasars and the expansion of the universe

(1) MIXING REDSHIFTS

Many hypotheses have been put forward to account for the large redshift of quasars. Some theorists have reasoned that they are relatively local objects and have ascribed the stretching of their wavelengths to something other than the expansion of the universe. Others have considered quasars as essentially cosmological, so that the redshift is attributed to the rapid expansion of the universe at great distances. And yet more astronomers have long suspected that there could be a mixture of effects, with a big slice of the redshift coming from a cosmological recession velocity, but a further contribution being added by gravitation or random velocities. In such a situation it is perhaps best to keep an open mind and search for critical tests of all the possible hypotheses.

Suppose an observer assigns a redshift z(obs) to a particular quasar. In principle this redshift could be due to recession velocity, random space velocity, and an intrinsic cause such as gravitation. Let us write these contributions respectively thus: z(Hubble), z(chaos) and z(internal). Then the following expression relates the different redshifts:

$$1 + z(\mathrm{obs}) = [1 + z(\mathrm{Hubble})]\,[1 + z(\mathrm{chaos})]\,[1 + z(\mathrm{internal})].$$

For the sake of argument suppose that z(obs) = 1.95. Then if the quasar is at a distance at which galaxies would have a redshift of only 1.00, the random motion and internal redshift mechanism would each have to contribute only about $z = 0.2$ in order to increase z(obs) to 1.95. Such effects would not necessarily be easy to detect, for the random motion would be only one-third the recession velocity in the case specified here.

(2) LOCAL THEORIES

The impetus behind the so-called local theories is the desire to construct a model of a quasar that does not require the mighty energy source needed if they are at cosmological distances. A suggestion dating from 1965, and first put forward by J. Terrell, was that quasars have been expelled from our Galaxy or a member of the local group of galaxies by a violent explosion. What we are now seeing is the ejecta of this event travelling away from us at close to the speed of light.

One powerful argument against this local hypothesis is basically philosophical: it is unreasonable for us to be located in a privileged position in the universe. If only our Galaxy has shot out quasar debris then we would indeed be in a unique place. To avoid this difficulty we suppose instead that quasar expulsion is not uncommon in galaxies, so that we are no longer a special case. Then we should expect to see a few quasars with large *blueshifts*, due to the quasars ejected from nearby galaxies towards us. In fact, the blueshift also makes quasars brighter at optical and radio wavelengths, so that samples selected on a brightness criterion might show more blueshifts than redshifts! No object with a large blueshift has ever been found and there is no class of objects with inexplicable spectra that could be a population of high blueshift quasars.

A second and more powerful argument against the local theories is provided by the radio astronomers. Besides the radio emission from discrete objects they also detect a uniform background of radio waves from the whole sky. If other large galaxies possessed an entourage of local quasars—as our Galaxy is supposed to according to the local hypothesis—then the contribution to the background from the myriads of quasi-stellar radio sources would probably make the radio sky much brighter than is in fact the case.

Further evidence against a local theory comes from the radio maps. Many of the quasars show the double structure that is already familiar for the radio galaxies. It is difficult to understand why local quasars and distant galaxies should have the same apparent radio structures as observed from Earth. This particular argument does not, however, go against a local hypothesis for the compact quasars which do not exhibit the double structure.

Another barrier to the widespread acceptance of the local hypothesis is that it does not necessarily diminish the energy problem. If a multitude of bright objects is to be banished from a galaxy at close to the speed of light, a not inconsiderable amount of kinetic energy has to be conjured up.

(3) GRAVITATIONAL REDSHIFTS

A quite different sequence of ideas attempts to use gravitation as the whole or partial means towards a large redshift. If electromagnetic radiation is generated in a region of high gravitational potential and observed in a region of lower gravitational potential, the observer will see a spectrum that is displaced towards lower wavelengths, i.e. a redshifted spectrum. A high gravitational field exists close to very massive compact objects, whereas the gravitation on the Earth's surface—which is where the telescopes are—is low. So light emitted from an object with a high gravitational attraction is observed to be redshifted. This effect is even detectable in the lines of the solar spectrum. The gravitational force at the sun's surface is twenty-eight times higher than on Earth; this field gives the same shift to the red as a recession velocity of 0.64 km/sec.

For a body of radius r and mass M the gravitational redshift of light emitted at its surface and received at the Earth is:

$$z(\text{gravitation}) = GM/rc^2,$$

where G is the gravitational constant and c the speed of light. We see that massive objects (large M) that are compact (small r) can have an appreciable gravitational shift. However, there are objections to making the gravitational redshift much greater than 0.5: structures with gravitational fields so intense that z exceeds 0.5 are unable to prevent themselves being crushed by their own forces. So the objects will probably collapse and in doing so decrease their radius. This only intensifies the gravitational forces, so that the collapse becomes catastrophic; an implosion would occur and probably a black hole be formed.

This argument is compelling but not watertight, because all it says in essence is that we have not succeeded in cooking up a model for high gravitational redshifts that does not destroy itself. Maybe the quasars are telling us that physics is cleverer than our mathematicians? However, it is true to say that if the redshifts are substantially gravitational (i.e. internal) then the counts of radio sources in the sky would probably be rather different, unless one again supposes that we are observing the universe from a privileged platform.

(4) TIRED LIGHT THEORIES

Another anti-Hubble-Law faction has cautioned against accepting that all the laws of physics are known. Historically great paradoxes in science have yielded to new and simplified laws, the application of which has opened up great vistas of fruitful research. Basically, the argument is that a redshifting mechanism that we have not yet discovered is at work.

A recurring theme in ideas of this type is the 'tired light' hypothesis. This hypothesis has been thrown into battle in attempts to decimate the vast armies of astronomers who believe in expanding universes. The fundamental proposal is that electromagnetic radiation, when it travels vast distances, loses energy, so that the photons get tired out. This cosmic diminution in their energy makes their wavelength longer, and so the spectrum of a distant object is redshifted. But where does the energy go to as the light tires? Suppose that the individual packets of energy—photons as they are called by physicists—react with other particles, say, electrons or atoms. They transfer their energy in collisions or interactions of some kind. But in doing so they will also lose information on where they are heading. The interaction will react back onto the photon, changing its direction of motion. Observationally, we would see images smeared out. A faraway quasar would have been seen as a disc not as a starlike image. This still leaves the possibility that the light participates in a mysterious self-interaction, enabling it to give up energy progressively as it goes, without having to feed it across to material particles. However, all processes in physics are essentially of a statistical nature. They involve uncertainties and a degree of randomness. The net result of this is that all photons starting out with the same energy could not contrive to lose it at exactly the same rate. A broadening in the spectral lines would therefore result; as the broadening is far less than the redshifts observed the photon-drag theory appears untenable. The Soviet theorist, Ya. B. Zel'dovich, has made known a general argument which shows that if the light does decay in energy, then the lower frequency waves (i.e. radio waves) would tire the most readily. If this happens one is soon faced with a maelstrom of deadly snags to general electromagnetic theory, which is well tried and tested. On balance it appears that notions of energy fatigue offer no attractive solution to the mystery of quasar redshifts.

(5) DISCREPANT REDSHIFTS

Of all observers none has pressed the theorists harder to keep an open mind than Halton Arp. He is one of a small number of independent thinkers who have set out to demonstrate that the Hubble Law is not all that it is cracked up to be. We have already seen in Chapter 6 that Arp and others have found examples of interacting galaxy clusters in which it appears that individual members may violate the Hubble relationship.

Arp showed that the quasar PHL 1226 is only 30 arc seconds away from the galaxy IC 1746 (Plate 22); the chance of an angular separation of this order between a quasar and a bright galaxy is claimed to be small. Again for the galaxy NGC 7603 Arp obtains a redshift of 0.027; but the galaxy apparently has an appendage with $z = 0.063$. Burbidge, Burbidge, Solomon and Stritmatter discovered that four of the quasars in the 3C catalogue lie within a few arc minutes of bright galaxies. Again the probability that this is a configuration which has arisen accidentally is said to be low—less than 1 per cent. Evidence of this type, leaning as it does so heavily on statistical arguments may be said to be suspect if it is to be used as ammunition against the acceptance of Hubble's Law. Nevertheless the possibility that these strange juxtapositions are genuine associations should not be dismissed out of hand. However, we shall not consider that they are sufficiently well founded to destroy the Hubble Law. The observational evidence favouring new laws is slim.

Sir Fred Hoyle has busily promoted a new law approach, not just to quasars, but to the more far-reaching problem of defining the gross properties of the universe. He has pointed out that if the masses of the fundamental particles and the atoms change with time, then the radiation from distant atoms will appear redshifted compared with radiation from similar atoms in the observer's laboratory. This idea is so radical that a detailed discussion is inappropriate here. It is another wedge we can use to keep our minds open to other solutions.

Except for a few chinks here and there the hatches seem to be battened down on non-cosmological theories of quasar redshift. By process of elimination we have reached a stage where the cosmological point of view—that quasars obey Hubble's Law—prevails. Faith in Hubble's Law is firmly backed by Sandage's observations for the bright cluster galaxies. Any other notion involves a mountain of rethinking, plus the overthrow of much theoretical work that has proved fruitful up to now.

(6) CONVENTIONAL VIEWPOINTS

James Gunn of the Hale Observatories came up with powerful evidence that some quasars at least have cosmological redshifts. He, and other workers since, examined quasars that lie in the same direction as clusters of galaxies. Hardly anyone disputes that the redshifts of galaxy clusters are due to the expansion of the universe. Gunn discovered that quasar PKS 2251 + 11, which lies on the same line of sight as a distant cluster, has the same redshift (0.33) as brighter cluster members. About half a dozen similar examples are known of quasars associated with clusters and having redshifts that agree closely.

Even in the case of clusters there is room for disquiet, however. Galaxy clusters cannot be discerned, even with the world's largest telescopes, beyond redshifts of about 0.6. So the Hubble Law for those quasars which are associated with clusters can only be confirmed for low redshifts. What hope is there of checking out redshifts of 2 or 3? To attempt just such a check, Cyril Hazard of Cambridge and his collaborators in Australia and the U.S.A. have looked into the authenticity of higher redshifts. They tried to set about their investigation in such a way as to avoid bias towards finding any preconceived results. A general search for quasars in association with galaxies brought a handful of interesting candidates to light.

Quasar 4C 24.23 is a blue compact object that nestles at the edge of a neat group of five galaxies that are probably the brightest members of a remote cluster. The redshift of 4C 24.23 is 1.27. In another case a faint galaxy cluster surrounds the object 4C 11.45; this has a redshift 2.17, so it must be a quasi-stellar object. Another radio source in the 4C catalogue, 26.48 has a redshift of 0.78, and it lies within 10 seconds of arc of a faint galaxy. In each case cited here the positions of the radio sources have been measured so accurately that there can be no significant doubt about the reliability of the identifications. Note, however, that galaxies become mere smudges on a photographic plate for redshifts exceeding about 0.3, and they are certainly invisible at redshifts greater than 0.6. So although none of the galaxies apparently near to the quasars has had a spectrum taken, we can be fairly certain that, if spectra were obtained, there would be a discrepancy between the galactic and quasar redshifts. All the alignments could be chance superpositions of objects at different distances, of course. The point we are making here is that the investigation of quasars and clusters of galaxies has only confirmed the cosmological nature of the redshift for relatively low values of redshift.

A profound mystery is presented by 4C 11.50, also the subject of an investigation by Hazard and his colleagues. For this radio source they found not one, but two quasars in the field of view. Both have since been confirmed as independent radio sources, one of them exhibiting the classical double structure. But radio maps released in Cambridge in 1974 showed no sign of interaction between the two quasars. A spectroscopic study at California's Lick Observatory found redshifts of 0.44 for the brighter—and presumably nearer—object, and of 1.90 for the fainter one. Curiously enough these values mean that the lines in one quasar are found at almost twice the wavelength in the other, since the wavelengths change as $(1 + z)$, and 2.90/1.44 is approximately 2. Is this just another unfortunate coincidence calculated to send the theorist up a blind alley? The quasar pair is separated on the sky by less than 5 seconds of arc. Observer Alan Stockton, working in Hawaii, has obtained image-tube photographs of the 4C 11.50 quasars and the surrounding field. These pictures penetrated to very faint magnitudes and recorded a necklace of fuzzy galaxies around the quasars. One of his image-tube pictures is reproduced as Plate 21. For one of the galaxies a redshift of 0.43 has been determined. This confirms a cosmological interpretation for the quasar with the lower redshift, but it still leaves the higher redshift open to question.

Perhaps a solution to the redshift controversy is now in sight. Radio telescopes capable of making fast surveys of broad swathes of sky are now in operation, and they will greatly increase the number of radio sources for which accurate positional measurements are available. With the establishment of large optical telescopes in Chile and Australia the opportunities to investigate southern skies with the same thoroughness as the northern heavens will increase. Within a few years the number of catalogued quasars should have increased considerably, and when that stage is reached it will be possible to place more confidence in arguments that are basically statistical. Extragalactic astronomy, however, is not an endeavour in which the so-called answers should ever be regarded as final and immutable!

11 · What lies between the galaxies?

11.1 Do we need intergalactic matter?

The profound mysteries of the origin, evolution and ultimate destiny of the universe depend to a large extent on how much matter the universe contains. We have already seen that the galaxies are moving away from each other; the further our telescopes probe, the faster these galaxies appear to be receding. Our observations indicate, therefore, that the universe of galaxies is in a state of expansion and that the mean separation of galaxies appears to be increasing with time. Clearly, it is of outstanding interest to determine whether the present expansion will continue for ever, or whether it will gradually slow down to zero velocity, or whether the expansion will eventually cease completely and then be replaced by contraction of the universe.

The ultimate fate depends on the balance between kinetic and gravitational energy. If the kinetic energy vested in the velocities of the expanding swarms of galaxies exceeds the gravitational energy of the mutual attractions, then gravity cannot restrain the expansion, even given infinite time. On the other hand, if the gravitational energy is larger than kinetic energy, gravity will eventually slow the expansion to zero, and then go on to cause the universe to start contracting. The singular condition where the two energies match exactly, and it takes an infinite time to halt the expansion, is probably of mathematical interest only.

The rate at which the universe is expanding is revealed to us through Hubble's constant H, which specifies the systematic velocity of a given galaxy at a given distance. In simple cosmologies it is possible to calculate a value for the critical mean density of matter which the universe must have if the expansion is to be halted ultimately by gravity. This critical density ρ(critical) is given by:

$$\rho(\text{critical}) = \frac{3H^2}{8\pi G}$$

where G is the universal gravitational constant, which specifies the magnitude of gravitational forces. If the average density of the universe is bigger than ρ(critical) the universe is said to be closed; its ultimate extent is then finite. If the average density is below ρ(critical) the universe is open and the mean spacing between galaxies therefore increases indefinitely.

This argument shows that the ultimate fate of the cosmos is intimately linked to the total amount of matter stored inside it. The observed values of H (100 km/sec/Mpc or 3.2×10^{-18} m/sec/m) and G (6.67×10^{-11} N/m²/kg) yield: $\rho(\text{critical}) = 2 \times 10^{-26}$ kg/m³.

The most conspicuous form in which we see matter in the universe at large is in the form of galaxies. Many astronomers have tried to compute the contribution to the mean density that comes from the material locked up in the galaxies. This can be worked out by first deriving the masses of different types of galaxies to find the mass distribution among the various types, and then making sample counts of the *numerical* density of galaxies in typical regions of deep space. By combining the masses and number densities of galaxies it is then possible to calculate the contribution to the total amount of matter made by visible galaxies. This was tried in 1971 by Thomas Noonan. He found that if the mass in galaxies were to be uniformly spread out it would contribute a universal density

$$\rho(\text{galaxies}) = 3 \times 10^{-28}\ \text{kg/m}^3$$

for H = 100 km/sec/Mpc. If we express this as a percentage it turns out that the *visible galaxies contribute only 1.5 per cent of the matter required to close the universe.* This percentage is independent of the value of Hubble's constant, the only uncertain factor in the expression for ρ(critical). Earlier investigations by J. Oort, for example, have yielded results similar to Noonan's.

The low value for the density of galaxies raises an important question: has the counting procedure missed vast hoards of faint galaxies that could, in fact, make a hefty contribution? Jim Peebles and Bruce Partridge have looked at this by considering what would happen observationally if the universe contains myriads of normal stars that are not in bright galaxies; they might, for example, be in undetectable dwarf galaxies or even intergalactic space. But wherever they are they will shine out, and the total effect of their stars would be to increase the brightness of the sky at night. The observed background light of the night sky implies that less than 13 per cent of the critical mass is in the form of normal main-sequence stars that are not in bright galaxies. Again, this percentage is independent of the value of H.

This, then, leads to a discussion of subluminous objects. Is the universal density inflated to the value needed to ensure closure by the addition of very faint stars (white and red dwarfs), black holes, rocks, stones or what? This question can be made more graphic by considering that the critical density is equivalent to one ordinary house brick in an otherwise empty cube of side 500,000 km! In principle, therefore, the mass of invisible rubbish could far exceed the critical mass. However, independent evidence can be brought to show that the real density of the universe cannot greatly exceed the magic density without drastically altering the appearance of far-off galaxies. For aesthetic and astrophysical reasons it has generally been supposed that the universe is, in fact, closed, and that the missing mass is in the form of a gas.

The question of the existence of the intergalactic medium is heavily dependent upon the masses of galaxies. The principal reason for postulating its existence is that galaxies alone do not provide enough mass to close the universe. However, there is the possibility that the masses of galaxies are themselves seriously in error. If galactic masses have been underestimated by a factor of 10 or so, the cosmological impetus behind the search for the intergalactic medium would cease.

In 1974 three Estonian astronomers, Jan Einasto, Ants Kaasik and Enn Saar claimed that the galactic masses are indeed too low by factors of about 10. To sustain the claim they gathered more precise data on the mass distribution in the peripheries of galaxies by observing bright objects moving in near circular orbits close to the plane. Five galaxies were examined in detail, and in each one the mass distribution attributable to the observable stars differed substantially from the mass distribution inferred analytically from dynamical considerations. They concluded that the galaxies contained a massive invisible population of matter. Other researchers have independently suggested that galaxies possess massive faint haloes. Their argument is that a massive halo will help to stabilize the central bar in the barred spiral galaxies against dynamical disruption.

It is possible that the binary-pair method of finding galactic masses is in error. One important effect is that observers more readily select pairs of galaxies that have only a small separation. As the separation between a pair increases the likelihood that it will, in fact, be recognized as a pair decreases. In a given case of a binary galaxy, the velocities along the line of sight become gradually smaller as the apparent separation decreases. Because the galactic time scales are so long there is no possibility of determining motions across the line of sight, and therefore only the radial velocities can be used. The data on galactic couples therefore have to be treated statistically because there is not enough information for finding the masses in any one system. Now if the sample is biased in favour of close pairs, as seems likely, on the average the velocities will be too small, due to the effect mentioned above. This in turn leads to the mean mass being underestimated. The only way to overcome this serious systematic error due to selection effects is to search without bias for binary pairs. This is best accomplished by using a computer to search catalogues of galaxies for all pairs of objects separated by less than a predetermined amount. The genuine pairs among the combinations thus found are then recognized from observations of the redshifts; any couples with closely matched redshifts are probably gravitationally bound pairs. Preliminary results from unbiased approach to the binary method indicate that the masses may well be higher than had been previously supposed. However, the idea that galaxies may in fact be sufficiently massive to close the universe is not widely held.

11.2 Properties of the intergalactic gas

First we consider what the gas could consist of, bearing in mind that galaxies and stars are composed principally of hydrogen. From a theoretical standpoint we can be certain that the galaxies themselves condensed out of a gas mainly composed of hydrogen. It is likely then, that hydrogen is the major component of the gas. Theoretical pictures of the early universe (the big-bang model) also indicate that 25 per cent, by mass, of the intergalactic gas could be helium. Elements heavier than hydrogen and helium are synthesized in galactic and stellar explosions. The expectation is that the intergalactic medium will have negligible pollution of heavy elements, since they largely remain within the galaxies where they are synthesized.

Questions as to the temperature and density of the gas can be answered by spectroscopy and other techniques. If the gas is rich in cold hydrogen, for example, we would expect to see strong absorption in the light from a remote galaxy or quasar; this would happen because the photons of light at particular frequencies can transfer energy to the cold hydrogen by raising its electron to a higher energy orbit. If, on the other hand, the gas is very hot, then we might anticipate strong X-rays from it. These are just two examples of the experiments that have been tried in the hunt for missing matter. The techniques used in the search depend on whether the gas is neutral (each atom possessing its full complement of orbital electrons) or ionized (atoms are excited and have lost one or more electrons), and these will be described separately here.

11.3 The hunt for neutral gas

A smooth distribution of cold hydrogen atoms at the critical density would be detectable by present techniques. The first attempts at this were made by radio astronomers. Cold hydrogen strongly absorbs at the rest wavelength of 21 cm, so that the radio emissions from powerful radio galaxies and quasars might be expected to show a sharp absorption line in their spectra. The intergalactic gas near a radio source would absorb, from our point of view, radiation at $21(1 + z)$ cm, where z is the redshift of the radio source. Hydrogen along the same line of sight but nearer us would absorb wavelengths between $21(1 + z)$ and our rest wavelength, 21 cm. So the radio spectrum of a source at redshift z would be expected to have a deficit between 21 and $21(1 + z)$ cm, caused by intergalactic hydrogen, if the substance were indeed present. The most accurate radio observations have failed to disclose a general pattern of 21-cm absorption bands in the spectra of bright radio sources. This places a limit of perhaps 20 per cent of ρ(critical) on the maximum contribution neutral hydrogen would make.

Far more stringent limits have been obtained by the optical astronomers, however. The fundamental transition in neutral hydrogen is the Lyman-α line at 1216 Å. Absorption occurs at this wavelength when an unexcited atom receives enough energy to release its electron from the ground state to the first excitation level. It is expected that neutral hydrogen in the depths of intergalactic space would absorb strongly at 1216 Å. This wavelength is in the ultraviolet, and consequently it cannot penetrate our atmosphere. But consider what happens to light from a quasar with, say, a redshift $z = 2$. Then Lyman-α is redshifted to $1216(1 + z)$ Å, or about 3600 Å, which is accessible to ground-based telescopes. In other words intergalactic neutral hydrogen would leave an absorption signature in the light from quasars at high redshift.

Jim Gunn and B. A. Peterson tried looking for a hydrogen absorption trough in quasar 3C 9, which has a redshift of $z = 2.01$. They, and later investigators, found no evidence for absorption. This fact leads to the result that less than one-millionth of the missing mass is in a uniformly distributed gas of neutral hydrogen. Strangely, this limit is almost too low! It implies either that the whole concept of quasar redshifts is false, or that galaxy formation is unbelievably efficient in sweeping up hydrogen atoms from neighbouring intergalactic space, or that what neutral hydrogen there is exists in denser clumps. Of these proposals, the clumpiness idea has received the strongest support. Another possibility is that the hydrogen atoms have all paired off to make hydrogen molecules. Observations, however, do not lend much support to this idea, as the expected absorption effects have escaped detection.

The main conclusion from spectroscopy is that any intergalactic medium which does exist will not be found to contain significant amounts of cold, neutral hydrogen. This forces us to think seriously on the possibility that the hydrogen atoms have been ionized and now consist of a plasma of unshackled protons and electrons.

11.4 A hot intergalactic medium

Observations have compelled us to conclude that if the hydrogen exists in intergalactic space in anything like the amounts needed to close the universe, then it must be ionized. This in turn implies that the gas must be very hot; certainly 100,000 K or more. The reason for this high temperature is that a colder gas would tend to recombine into neutral atoms too quickly, and in any case plausible mechanisms for producing the ionization will actually heat up the gas as well. The production of a high temperature in intergalactic space is not of itself a problem. We could suppose, for example, that when the radio galaxies and quasars first formed they poured vast amounts of ultraviolet radiation into the intergalactic medium, and it would then have been possible for them to heat the gas to a temperature of perhaps 10^6 K.

The most direct information on the existence of ionized gas comes from observations of cosmic X-rays, because a hot ionized gas is a significant source of thermal radiation in the X-ray region of the spectrum. The observed X-ray intensity gives an upper limit of 3×10^8 K for the temperature of an intergalactic gas at the magic density; the temperature most widely accepted in the literature is 10^6 K. At present the X-ray data do not eliminate the possibility that all the intergalactic medium is hot and ionized, but neither do they prove that it is there at all.

Substantial quantities of ionized gas will affect the observed brightness of sources of radiation at large distances. This is because electrons in the gas scatter the radiation out of the line of sight and so reduce the amount reaching the observer. Exact calculations show, however, that an ionized gas at the critical density gives a scattering effect that is smaller than intensity changes due to geometrical effects predicted by the theory of relativity. So measurements of the magnitudes of distant galaxies and quasars appear to be incapable of revealing the intergalactic medium.

11.5 Is the gas near to clusters of galaxies?

Up to this point we have assumed that the intergalactic gas is uniformly distributed. However, there is the very real possibility that significant concentrations will only be found in the vicinity of galaxy clusters. In 1962 Fritz Zwicky claimed that faint clusters of galaxies were less numerous within 1° of large nearby clusters. One interpretation of this is that intergalactic material close to the nearby clusters is absorbing light from more distant clusters, so that the apparent density of distant clusters is decreased. The difficulty in discussing this evidence is that the clusters themselves are widely believed to contain missing mass; only 12 per cent of the probable mass of the Coma cluster is in the form of visible galaxies. Is the other 88 per cent almost invisible gas in the general neighbourhood of the cluster? The Zwicky effect is thought to be incompatible with the distribution that the matter has to have if it is the missing matter binding the clusters. The conclusion, then, is that Zwicky's obscuration could be caused by intergalactic matter. Observations of the X-ray emission from rich galaxy clusters strongly suggest that they contain a very hot intracluster gas. However, George Field, among others, has expressed the view that if the hot gas in Coma is typical of all clusters, then cluster gas can account for only 1 per cent of the critical density. It is unlikely, therefore, that clusters contain the matter that would close the universe.

11.6 Electrons and magnetic fields in intergalactic space

Radio astronomy has provided information on the role of free electrons, such as would be found in an ionized gas, and large-scale magnetic fields in space. Many radio galaxies and quasars show a degree of linear polarization in their radio emissions. Measurements of the planes of polarization of radio sources at many different wavelengths indicate that Faraday rotation takes place somewhere in the space between the distant sources and our radio telescopes. The observed rotation need not necessarily be occurring in the intergalactic medium, of course, but if we assume that it is, we get limits on the strength of the magnetic field.

For the sake of argument we will assume that a fully ionized medium of the critical density exists. It then turns out that the Faraday rotation can be accounted for as an intergalactic effect if there are very-large-scale magnetic fields, and well-aligned ones at that, with a very small strength of about 10^{-12} webers or so. This is about one-hundred-millionth the strength of the Earth's magnetic field. If the Faraday rotation is not due to intergalactic effects, then the uniform field strength must be less than that stated, unless the medium is more dilute than we have assumed, in which case the magnetic field has to be stronger. One point that has to be borne in mind, however, is that Faraday rotation only reveals highly organized magnetic fields, because the uniform rotation requires a long path length under astronomical circumstances. The basic situation for magnetic fields, then, is that they must be very weak, unless free electrons are virtually absent in intergalactic space.

11.7 Snow, grit, bricks and planets?

We have already mentioned that a very thinly distributed population of bricks will close the universe. House bricks are implausible components of the universe, but what about other forms of cold dark matter that have been observed? Is intergalactic space filled with dust, meteorites and small planets which would be undetectable telescopically? Matter in such form could add enormously to the intergalactic mass density. When this consideration is probed fairly critically, however, it seems unlikely. For a start there is the observation that hydrogen is by far the most abundant element in the original composition of our Galaxy. Indeed, we have no strong reason for disbelieving that our Galaxy condensed from an intergalactic medium composed of hydrogen and helium alone. It therefore seems unreasonable that the intergalactic medium should now be composed of only non-volatiles whose total mass is substantially larger than that of all the galaxies. The argument then devolves to the possibility that the hydrogen has all turned into snow and ice.

The snow hypothesis can be subjected to test fortunately, as we can calculate the minimum size that the snowflakes have to have. If the snow were too fine, the light from distant galaxies would be attenuated. In order to picture the situation the following terrestrial analogue can be considered. Dense fog consists of extremely small water droplets and strong attenuation of light takes place for paths of a few metres. Rainfall, on the other hand, is made up of large water drops, and strong attenuation does not occur. In the intergalactic case, how fine is too fine? Galaxies out to redshifts of $z = 0.2$ do not show any detectable dimming which could be attributed to intergalactic haze. This leads to the conclusion that a distribution of snowflakes will only add up to the critical density if their radii exceed 1 cm.

By considering the stability of the hydrogen snowflakes we can derive even more stringent limits. The whole universe is bathed in background thermal radiation at a temperature of 2.7 K. This may seem very low, but it is not sufficient to permit indefinite deep-freezing of hydrogen. In this temperature bath the snowflakes evaporate quickly. They can only withstand this melting over cosmic time scales if they are rather large, about 10 km in size.

In conclusion, we can state that solid hydrogen can exist as a permanent inhabitant of intergalactic space only if it has formed hydrogen icebergs. Although we cannot disprove that these exist, it is difficult to understand how they could have formed.

11.8 Will-o'-the-wisps to close the universe

So far our consideration of the total mass of the universe has taken in tangible aggregations such as gases, plasma, stars and galaxies.

According to the general theory of relativity, energy E and mass m are related thus:

$$E = mc^2,$$

where c is the velocity of light. This states that the energy equivalent of a mass m is mc^2, and conversely, we must assign to an energy E a mass E/c^2. The implication of the relation is that the total mass should take into account the mass contributions locked into electromagnetic radiation. The great strides made by instrument developers mean that we can sample the cosmic electromagnetic radiation from long radio waves right through to high-energy gamma-rays, across 16 orders of magnitude in frequency range. When the spectrum of the universal background radiations is plotted, it turns out to contribute much less than the critical mass to the mean density of the universe. Under the electromagnetic heading we must also include the mass-equivalent of the intergalactic magnetic field. As we saw above this is likely to be in the region of 10^{-12} webers, at which level it would yield a derisory mass-density.

There may be very-high-energy nuclear particles known as cosmic rays in intergalactic space. Currently, we have no means of finding out if this is so. However, galaxies do contain cosmic-ray particles and some of these must surely leak out into the interstices between the galaxies. Some astronomers have argued that cosmic rays are universal, in which case their possible mass contribution needs consideration. Since we know nothing about extragalactic cosmic rays we could assume that the extragalactic density is not likely to exceed that in our Galaxy, in which case cosmic rays account for only 0.01 per cent of the critical matter of density.

Among the most esoteric particles in the whole of physics is the neutrino, a chargeless bundle of energy with zero rest-mass. Because neutrinos have almost zero interaction cross-section with ordinary matter they are next to impossible to detect. In a South Dakota gold mine Ray Davis has spent many years trying to detect neutrinos from the sun, but he has detected very few. The day when neutrino telescopes will scan the galaxies still seems far in the future. Consequently, it is not physically impossible that vast amounts of missing mass, sufficient to close the universe many times over, is vested in these elusive will-o'-the-wisps.

Finally, we can round off our census of the inhabitants of the universe's voids with black holes. All that can be said here is that we must recognize that black holes could provide all of the missing mass. If these were distributed fairly uniformly, and were not too large, we could not detect them. All we can fall back on is the weak argument that it is hard to see how vast amounts of matter could imprison themselves in these bottomless gravitational wells without some of the side effects being noticeable.

11.9 Concluding remarks

The systematic study of the properties of the intergalactic medium provides an object lesson in scientific reasoning. Despite our inability to detect a universal medium we can place valuable constraints on some of its physical properties purely because certain effects are not seen. If it is there it has to obey certain rules in order to remain hidden. Of all the physical sciences astronomy, perhaps the most frequently, has to fall back on arguments that are based on an absence of information. This is largely because its practitioners often cannot set up laboratory experiments for probing the properties of astrophysical materials.

A fundamental cause for concern arising from investigations of the intergalactic medium is the possibility that only a small fraction—less than 20 per cent—of the mass of the universe may be contained in normal and dwarf galaxies. There is also the profound problem of the total mass of the universe. How can we arrive at a picture of the total contents of the universe if we do not even know what fraction of the whole each of the observable forms occupies? Since the sixteenth century science has been in retreat over the relative importance, cosmologically, of different celestial bodies. The sun replaced Earth as the focus of the heavens with the Galilean revolution. The eighteenth and nineteenth centuries saw the sun slip in prominence, until it was positioned insignificantly at the rim of the Milky Way early this century. Hubble cracked the enigma of the nebulae and showed that our Galaxy is but one of innumerable myriads of galaxies. Can we be sure that the great clusters of galaxies are the most important component of the universe? We are even today not at the end of accumulating an elementary picture of the true contents of the universe, although we can paint parts of the picture in exquisite detail.

12 · The energy problem

12.1 Trouble with radio galaxies

Observatories with radio telescopes have shown that in many radio galaxies emission spreads over regions of many hundreds of kiloparsecs. In the two giants 3C 236 and DA 240 the size even runs into several megaparsecs. Within these radio clouds, electromagnetic radiation is generated through the process of synchrotron emission. This radiation mechanism requires two basic ingredients: electrons travelling at highly relativistic velocities, so that their energy is much greater than the energy at rest, and magnetic field. These two components, fast particles and magnetic field, will be referred to as relativistic energy in this discussion.

One of the early questions that was tackled when the sizes of radio galaxies had been determined was the total relativistic energy required to maintain the radiation. Of course, in order to work out sizes, distances derived from redshifts are used, but in the case of radio galaxies the reliability of redshifts as a distance indicator is generally accepted. Once the volume filled by relativistic energy has been measured observationally, the energy needed to maintain the emission over the lifetime of the source can be computed.

Standard equations in electromagnetic theory describe the radiation emitted by a particle that is trapped in a magnetic field. However, in order to apply the theory to radio galaxies, the relative proportions of energy stored as fast electrons or magnetic field must be decided. Unfortunately, the ratio of these two energies cannot be found observationally. This does not mean all is lost, however, because what is done in practice is to assume that the total energy needed to power the radio source is as small as possible. It turns out that this minimum situation holds if roughly half the available energy is used to make high-energy electrons, with the other half being utilized in creating the huge magnetic field system which threads through the radio components. It might be argued that Nature is not at all likely to create radio sources in a way that suits our inability to measure the ratio of energy for ourselves. Nevertheless the point is that the assumption is the most conservative one possible. In practice the energies will be higher than those obtained from this equipartition.

Another observational problem arises because the radio emission detected only comes from electrons, which have a negative electrical charge. Our general experience is that sufficiently large volumes of space are essentially electrically neutral. Presumably the electrons have come from hydrogen atoms that have been ionized. It is therefore likely that the radio components contain as many protons as electrons. A decision must now be made on how much energy should be assigned to the protons, in comparison to the electrons. The only fact that can be brought to bear on this imponderable is that in the cosmic radiation protons possess about 100 times as much energy as electrons. Whether this is of any relevance to the situation obtaining inside radio sources cannot be said. However, if it is assumed that the protons are 100 times more energetic then the minimum energy is increased by a factor of about 10.

When the minimum energy is calculated for the largest and brightest extragalactic sources, an energy requirement of 10^{51}—10^{53} joules is necessary to account for the observations. In the smaller sources values in a range 10^{49}–10^{51} joules are obtained. On the most conservative possible assumptions we would need 10^{52} joules to create a radio source similar to Cygnus A. Since we are here talking only about radio sources which are identified with giant elliptical galaxies, the energy requirements cannot be significantly reduced by postulating that the distance estimates are in serious error.

It is instructive at this stage to consider the rest-mass energy of the sun, in order to put the radio source energies into perspective. The rest-mass energy is computed by multiplying the sun's mass (2×10^{30} kg) by square of the speed of light (9×10^{16} m^2 sec^{-2}), and it comes to 1.8×10^{47} joules. Physically this is the energy that would be released if the mass of the sun could be converted entirely into energy. In practice the nuclear reactions going on inside stars convert less than 1 per cent of the star's matter into energy, leaving a massive remnant, such as a white dwarf, at the end of the evolution. So the astrophysically useful energy given out by a star like the sun is about 1.8×10^{45} joules.

Now we can compare these two energy parameters for the sun with our recipe for manufacturing radio galaxies. If we are allowed to turn matter into energy with total efficiency we need something like ($10^{52}/10^{47}$) solar masses, i.e. 100,000 stars. On the other hand, if we are only allowed to use conventional astrophysics, 10 million solar masses must be involved in the production of the requisite energy. The energy problem for extragalactic radio sources is the problem of accounting for the prodigious amount of energy, and then turning it into relativistic particles and magnetic field.

12.2 Conventional energy sources

It is observationally certain that the energy production mechanism, no matter what it is, is associated with the nuclei of galaxies. It is not yet known whether the energy is all released in a comparatively short time (say 1000 years), or whether it is produced continuously or intermittently throughout the life of the radio source (probably 1 million years or more). Consequently, both possibilities may be considered at present.

Perhaps the theories that stick closest to well-trodden ground are those involving supernova explosions. In the discs of spiral galaxies like our own the supernova rate is roughly one every fifty years. Because the sun is situated in the plane of the Milky Way we do not see as many as this on account of interstellar obscuration, and none has been seen since the invention of the optical telescope. In a powerful explosion perhaps 10^{43} joules of energy will be released by nuclear processes that we can make some claim to understand. This amount is sufficient to maintain the energy demands of the weaker radio sources, provided that several supernovae go off each year in the nucleus. In the brighter compact radio sources something like a supernova a week is necessary. Why should the supernova rate in a galactic nucleus be so much higher than in the disc? The answer is that the gas density and star density are much higher in the nucleus, so the rate of star formation is greatly enhanced. The supernova idea is perhaps more tricky if it is desired to produce all the energy at once, because then a means of triggering star formation at an extremely high rate has to be suggested. Some millions of solar masses of material have to be induced to form massive stars that evolve rapidly.

As a side issue to the basic supernova hypothesis, the idea that the relics of the explosion might play a part in continuous energy generation has been set out. It is believed that supernova explosions will lead to the formation of extremely dense, spinning neutron stars. Such objects have already been detected as pulsars within the Galaxy. These dense spinning relics possess considerable energy by virtue of their high angular momentum. Physicists have already worked out, for the case of pulsars, how this energy might be used to accelerate particles to high velocities, and it seems that this can be done quite efficiently. Another suggestion is that pulsars sitting in the centre of a nucleus might accelerate beams of particles, which are then channelled out through the galaxy, across deep space to the pair of radio components.

Like most theories of energy generation, the supernova hypothesis and its variants are practically impossible to subject to direct observational test. Even the brightest supernovae are difficult to detect if they are further away than the Virgo Cluster. Secondly, a search for supernovae is very much a hit-and-miss affair, and is not therefore likely to be popular with the project committees of large telescopes. Finally, there is a host of ways in which the exploding supernovae could remain hidden from view if they are going off in a dense nucleus.

A different idea involving the stars in the nucleus has been pursued by Stirling Colgate and his colleagues. They envisage a galactic nucleus with a star density so high that collisions between individual pairs take place frequently. If the star density is high enough, the velocities in the nuclear supercluster could run into hundreds of kilometres per second. Head-on and near head-on collisions then involve substantial kinetic energy and this can be channelled away from the stars if they shed their outer layers. In this way a store of highly excited gas is built up. In this model the source of energy is the gravitational field within the cluster, because continual collisions will cause the cluster to contract, with the field becoming gradually more intense. As in the case of the supernova hypothesis, it is difficult to imagine how the model could be tested observationally, at least in a way that would enable it to be distinguishable from its rivals. The supernova and stellar collision ideas can be fused together by postulating that the collisions cause larger stars to be assembled which then become supernovae.

The realm of stellar models is extended further with the concept of supermassive rotators at the nuclei of active galaxies. Extremely massive stars, of say 100,000 solar masses, are thought by some theorists to be stable provided they are set in rotation, in which event centrifugal forces can counteract gravitation. A spinning massive object, called a spinar by some workers, would store a vast amount of energy by virtue of its angular momentum. The concept enjoyed popularity when it was believed that some quasars exhibited quasi-periodic variations, because it was thought that the rotation could account for the semi-regular behaviour. However, subsequent observations have tended to discredit the claims that the variations are other than random, and the idea has fallen into disfavour. If large numbers of stars provide the energy, however, it must be borne in mind that the end-product of stellar collisions and supernovae is a vast quantity—perhaps 10 million solar masses or more—of stellar clinker. The ashes will remain in the nucleus and agglomerate on account of mutual gravitational attraction. In this way a huge garbage heap, or supermassive object, comes into being, although it is not at all clear whether or not it could make any further useful contributions to energy production.

12.3 Black holes and white holes

Dissatisfaction with the lack of success of stellar astrophysics to explain the energy requirements of radio sources led to the exploration of more speculative models, especially those involving black holes. The advantage of a black hole theory is that the energy is gravitational in origin and it is therefore easier, in principle at any rate, to convert a reasonable proportion of the mass into energy than is the case with nuclear reactions inside the stars.

To see what can be done with the black holes we start off by considering a conventional celestial object that is familiar to all of us, the Earth. We are going to conduct a 'thought experiment' with the Earth. As is well known the Earth has an escape velocity of about 11 km/sec. An object which attains this speed can escape the Earth without need of any further energy input. This is also the speed at which an object released at rest from a great distance would hit the top of the Earth's atmosphere. Now let us work out how much kinetic energy is brought to Earth by an object of mass 1 kilogram which is allowed to fall on Earth from deep space. At impact it is travelling at roughly 10,000 m/sec (the escape velocity). Its kinetic energy is given by $\frac{1}{2}mv^2$, and for the mass chosen 0.5×10^8 joules of energy are released—sufficient to keep a 1 kw heater running for a week, but an infinitesimal fraction of a radio galaxy's requirements.

Obviously the energy released by dumping interstellar matter onto the Earth can be increased by sending more of it towards Earth. If sufficient matter rained down, the energy released would heat up the atmosphere until it boiled off, and heat up the surface. But we still could not make a radio galaxy, not even by dumping an amount of rubbish comparable to Earth's own mass. Nonetheless, we have learned something from the thought experiment, because each kilogram brings with it 50 million joules of energy, and this energy has not been created inside the stars. Instead it has been extracted gravitationally. After each kilogram reaches Earth, Earth's gravitational field has been very slightly increased; the Earth is sitting in a deeper 'potential well'. What this means is that as matter accretes onto the Earth, intensifying the gravitation, it would require more and more energy to take the Earth to pieces and disperse all the bits far out in space. However, in this thought experiment we do not envisage ever wanting to dismantle the Earth, and so we are pleased to be able to get gravitational energy at the expense of making it more difficult to demolish Earth.

Suppose we now let interstellar matter rain down on Earth at a very high rate, and go and view events from a planet round a nearby star. After a time we would notice that the infalling matter was coming in at a higher velocity. This is because the matter that has already arrived has increased the mass and hence the velocity of escape.

We continue by causing ever increasing quantities of matter to fall onto Earth. Soon the planet is more massive than Jupiter and interesting internal conditions arise. The central pressure and temperature become so high that nuclear reactions commence and we have turned the Earth into a star. We continue piling it with more mass until it is, say, twenty times more massive than the sun, and then we let it evolve as a normal star. Because this thought experiment is designed to show what can be done with gravitation we shall ignore the fact that many millions of years must elapse while our Earth-star burns up its fuel.

Eventually it has done so, and the energy source within the Earth-star is shut off. Now the gases that compose the star begin to cool, and so the internal pressure drops. This is followed by a contraction of the star and as the star contracts the escape velocity increases like

1/R, where R is the radius. As the contraction takes place the gravitational field intensifies, and so the star is squeezed even more intensely. A runaway collapse or implosion has started. At some stage a supernova explosion may occur, but even if it does we are still likely to be left with a tiny dense core of matter with a mass of a few solar masses. Let us continue our thought experiment, but consider only what we can do with this dense central relic.

Can anything stop the central core from collapsing ultimately? The answer appears to be no if the mass is greater than about two solar masses. No pressure or force known in physics can halt the inexorable plunge downwards to progressively smaller and denser configurations. Meanwhile, what is happening to the escape velocity? This grows larger and larger as the radius decreases and the energy brought in by each kilogram of infalling matter increases as well. Eventually, a stage will be reached when the escape velocity is equal to the speed of light. When that happens a *black hole* has been created. It is called a black hole because its gravitational field is so intense that nothing can get away from it, not even radiation, because the escape velocity exceeds the velocity of light. Consequently, nothing at all can be seen in the part of space occupied by the hole, and so that place will appear perfectly black. Matter falling into the black hole will release a large fraction of its rest-mass energy, which presumably goes into heating the hole.

The picture given above is intended solely as an illustration, because black holes are not actually made by accreting matter onto planets. However, there are reasons for believing that they might be formed in supernovae explosions, or in galactic nuclei where the star density is high so that coalescence of stars is common. The black hole is an immensely useful theoretical device because it enables a large amount of energy to be released.

In our thought experiment we still have a problem if we wish to apply the general principle to the nuclei of radio galaxies, because the discussion has been in terms of energy being released when the matter hits the surface of the compact object. But, if the latter is a black hole, it is by then too late; any energy released at the surface of the body is useless because it remains trapped inside the warped space around the hole. Somehow we must so arrange the black hole that the accreting matter gives up plenty of its energy before plunging down into the bottomless pit. The answer possibly lies in rotation. If the black hole is given angular momentum, and if the matter falling in has angular momentum also, then almost half of the rest-mass energy can be extracted while the matter is still in contact with the outside world. If the infall material has angular momentum, individual particles go into orbit around the black hole rather than crashing into it. A spread in angular momentum values will give a population of particles moving on different orbits, so that if the accretion is high, a rotating disc of matter is swirling around the hole. Now we must add another ingredient to remove the angular momentum slowly, otherwise the particles will just orbit the hole without ever falling inside. Actually, friction, viscosity and collisions will all ensure that the matter progressively moves from the outer edge of the accretion disc to the inner edge, whence it falls down the hole. The accretion disc gets rid of the energy, thus extracted from the gravitational field, by radiation.

A scenario for the nuclei of active galaxies can now be sketched. At the centre of the galaxy is a massive black hole, which may have formed through coalescence of dead stars. Around the black hole is a large accretion disc. The disc is ravenously feeding on the interstellar gas and passing stars. If the material is processed through the disc in the ideal manner before being cast into the ever-greedy black hole, almost half of its rest energy can be obtained. This energy is then used up in the acceleration of particles to relativistic energies and in the creation of magnetic field. If the black hole—accretion disc combination operates with maximum efficiency, less than 100,000 solar masses of fuel has to be supplied in order to satisfy the minimum energy requirements of radio galaxies. However, as we have said before there is no guarantee whatever that Nature will perform for us at the theoretical efficiency and then make radio sources in the most economical fashion. So we might wish to add a factor of 100 or so for good measure. Even if we do this, it is only required to throw 10 million solar masses of material down the black hole. In the nucleus of a giant elliptical galaxy it would not be difficult to find this amount of mass. Furthermore, our present methods of determining the masses of galaxies are incapable of detecting a 10^7 solar-mass black hole at the centre of a 10^{12} solar-mass giant elliptical.

By postulating the existence of black holes of various masses in the nuclei of individual galaxies, it would be possible to account for the energy requirements of a whole range of nuclear activity. Martin Rees and Donald Lynden-Bell have, for example, put forward the idea that the nucleus of our Galaxy has a small black hole, in order to account for the large flux of infrared energy coming from the galactic centre.

An even more speculative idea is that of white holes. The argument runs that the nuclei of galaxies are places where new energy is being poured out into the universe spontaneously. No special machinery is needed for extracting energy from matter already in the universe. The white holes are postulated as sources of completely new energy, and this energy emerges in the nuclei of galaxies. One snag with the speculation is that the white hole might not survive if there is matter falling into it, because accretion is thought to turn white holes into black holes!

12.4 Taming the energy

Even if the energy source in active nuclei can be decided upon theoretically, there is still the question to be faced of creating extensive magnetic fields and a large supply of relativistic particles. In order to produce the observed duplicity of source components, some method of guiding the energetic particles in two opposite directions has to be thought out. From a theoretical standpoint, it may be easier to get the energy out in the right form if the process of energy release is continuous, rather than being a single mighty bang. Evidence favouring the idea of continuous energy release comes from maps of radio sources such as Cygnus A, in which a small amount of emission from a central radio component associated with the galaxy can be seen at high frequencies.

It must be emphasized that most of the remarks in this chapter are highly speculative, and within a few years it may have been shown that none of them work. But radio galaxies and quasars are important sources of energy, particles and magnetic field in extragalactic space, and some attempt must be made to understand them. It has been suggested that new laws of physics are at work, and until we have discovered what the new laws are we will not find out what makes active nuclei tick. Personally, I feel that we should not use this particular escape route until it has been plainly demonstrated that physics as we understand it is incapable of accounting for the observed energies.

13 · Origin and evolution of the galaxies

13.1 The beginning of time

Galaxies are evidently extremely ancient objects, and the question of their formation and evolution is deeply connected with the past history of the universe. The study of the origin, nature, and evolution of the universe is termed cosmology. One approach to cosmology is observational. In order to find out about the early evolution of galaxies and the universe we can look at the large redshift objects; the radiation from these galaxies and quasars has taken billions of years to reach Earth, and it gives us a snapshot of what the universe looked like long ago. Another approach to the question of the origin of the galaxies is the deductive theoretical treatment. The way this game is usually played is to take a plausible model for the universe, or cosmological model as it is called, and then see how and when galaxies might have formed in it. Earlier chapters have described the observations of high redshift objects, and these disclose the presence of very energetic phenomena in the early universe. In this chapter we shall outline results from the theoretical field, and in order to do this we first describe a critically important cosmological experiment: the discovery of the universal background radiation.

The first communications satellites, Echo and Telstar, helped to turn cosmology upside down in 1965. At Bell Laboratories, in Holmdel, New Jersey, communications is big business, and so a large steerable telescope was constructed for experiments with the satellites. While running their experiments, Arno Penzias and Robert Wilson became puzzled by a faint but perpetual hiss of microwaves from all parts of the sky, day and night. After sniffing around for sources of interference or possible error they could find no obvious local explanation for the anomalous signals, but like true physicists they still felt distrustful of their own results. Meanwhile, in another laboratory down the road at Princeton, physicist Robert Dicke and his colleagues were building a radiometer to see if they could pick up microwave radiation from the sky. What motivated this project was a 1948 prediction of George Gamow, developed later by Dicke, that the universe might have had a fiery birth in a stupendous big bang. Dicke argued that if the universe had been born in (or emerged from) a great explosion, then it might still be possible to detect a faint fossil signal of the initial violent event. When the New Jersey and Princeton groups contacted each other it was soon decided that the background radiation from the primeval fireball really had been detected as a persistent hiss in the Holmdel receiver.

A critical test of the fireball hypothesis was the measurement of the spectrum of the radiation. After one or two alarms and false starts it was eventually confirmed that the cosmic background radiation had the same spectrum as a black-body at 2.7 K (Figure 35). Another test that has been successfully completed is the isotropy criterion: if the radiation is the dying echo of the primeval big bang, then there should be no preferred direction in space, and consequently the temperature and spectrum of the radiation ought to be the same in all parts of the sky. This is indeed what has been observed in a series of spectacular measurements.

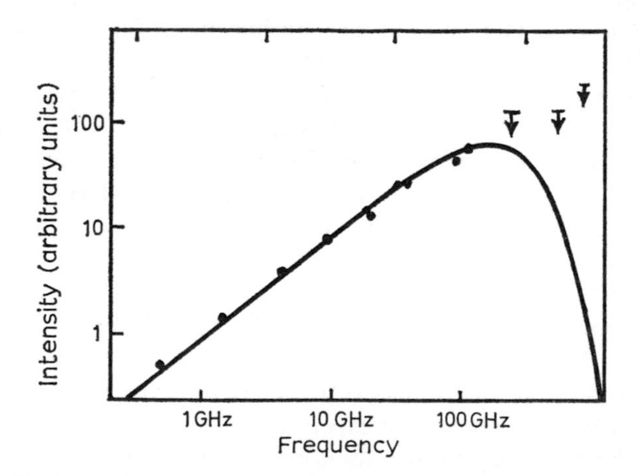

Fig. 35. The spectrum of the cosmic microwave background corresponds to the emission from a black body at 2.7 K. Upper limits have been obtained by observing CN and CH interstellar molecules

Partly as a result of the discovery of the background, practically all astronomers accept the basic idea of the big-bang model of the universe, which states that the universe evolved to its present state from a very hot, dense, and compact primeval atom, or singularity. The cosmic microwave radiation is a fossil remnant of this early hot phase, which is often incorrectly referred to as the origin of the universe. Expansion of the universe has subsequently stretched out the wavelength of the photons in the background radiation and thus inexorably lowered its temperature from billions of degrees down to a mere 2.7 K, so that it now peaks in emission at about 1 mm, with a tail out to a few centimetres. The radiation has the spectrum we would expect to see if we lived inside a perfectly black box with a temperature of 2.7 K; the black box is actually our observable universe. Its present mass-energy density is 4×10^{-37} kg/m^3, considerably less than the mean density of the luminous matter in the universe. From 1965 onwards extensive theoretical investigations have produced a rather detailed picture of how the universe has managed to get to its present state from the early fireball.

During the first second of time after the big bang the radiation had a temperature of a little more than 10^{10} K. Matter and radiation were probably in equilibrium, the matter consisting of a Greek alphabet soup of mesons, hyperons, neutrinos and positrons. Once things cooled down to 10^{10} K or so, the exotic high-energy particles started to combine and so make more familiar stuff like protons and electrons.

The next significant epoch in the early history of our universe was from 2–1000 seconds, when most of the primordial element synthesis occurred. Up until $t = 2$ seconds the particles were charging about so fast that any heavier nuclei that did form were soon smashed to splinters. After 1000 seconds the universe became relatively colder (10^9 K) so that the particles did not then have sufficient energy to overcome each other's electron repulsion during close encounters. Once the 1000 seconds was up the matter consisted of around 25 per cent by mass of helium nuclei, 75 per cent hydrogen atoms, with a trace of deuterium and lithium. After this early burst of element synthesis essentially nothing of great interest happened in the next 100,000 years, during which time radiation reigned supreme, its density being higher than that of the matter, and sufficient to keep everything ionized.

But all the time expansion was weakening the hold of radiation;

radiation density falls as the fourth power of the radius of the universe, but the matter density declines only as the third power, with the result that it gradually gained in relative importance. After 100,000 years matter overthrew the debilitated radiation dynasty: protons, nuclei and electrons joined forces to make atoms. This event was crucial to all subsequent evolution because once atoms had formed the radiation could no longer interact strongly with matter; a significant leverage only occurs when there are plenty of free charged particles that can lock onto the electromagnetic radiation, and once the nuclei and atoms have combined there are no longer large quantities of free charged particles.

13.2 Protogalaxies from perturbations

The mysteries surrounding the formation of galaxes are locked away down corridors of time that stretch back to the earliest history of the universe. The general way in which theorists play the galaxy formation game in the context of big-bang cosmologies is to postulate the existence of initial fluctuations in the density of matter, and then to explore the fate of these early perturbations as time advances. Expansion of the universe is continually reducing the mean density of matter. This means that unless density fluctuations are present in the early universe they never get time to grow into galaxies by gravitational forces. The nature of the primeval fluctuations is entirely conjectural, but it is usual to assume that processes in the fireball, prior to the recombination of matter, must have modified these density enhancements in such a way that concentrations massive enough to make galaxies and clusters of galaxies emerged. Galaxies would subsequently have formed by gravitational contraction out of these density irregularities.

Although the idea that galaxies may have formed from primeval irregularities is conceptionally simple, it is no more than a working hypothesis. We have no observational evidence that, on a galactic scale, the universe was ever more uniform than it is now. It is not, therefore, yet possible to account for the presence of structure in the universe, without judicious choice of initial conditions to fit the desired end.

The discovery of protogalaxies would give a valuable boost to the theoretical study of galaxy formation. It is expected that the protogalaxies formed so far back in time that we should have to look for

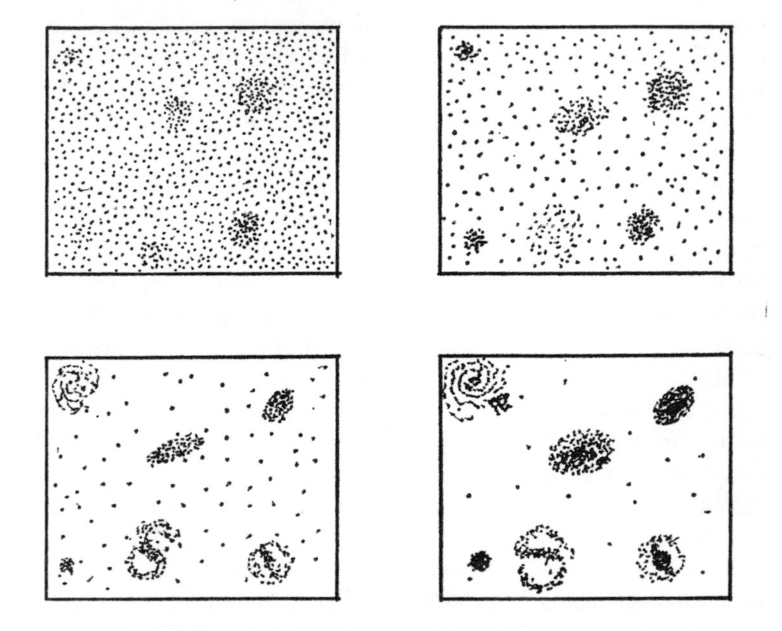

Fig. 36. Successive stages in a speculative scenario for the formation of galaxies from condensations in the early universe

high redshift objects. Since quasars are known with $z = 3.5$ it might be thought that these are protogalaxies. However, on balance, it seems more likely that quasars are an evolutionary phase in galaxies that have already formed. The upper boundary for the location of protogalaxies is perhaps at $z = 50$; until this point the universe has not existed long enough for the agglomeration of protogalactic clouds of the requisite mass and density.

Bruce Partridge has argued that young galaxies may need to divest themselves of a lot of energy quickly, and so be very luminous. Perhaps they would be detectable as far as $z = 10$. An optical search for fluctuations in the background light of the night sky is obviously of importance in the search for bright protogalaxies. So far the results of such experiments are inconclusive.

13.3 Evolution of galaxies

The problems surrounding galaxy formation are great, but granted that galaxies do exist, we are on comparatively safer ground when discussing their evolution. Evolution here can take on several meanings: the chemical evolution of the galactic material; the morphological evolution; and the dynamical evolution.

After commencement of star formation in a galaxy the chemical composition of its material changes. This is because the primordial material, from which the galaxies condensed, was composed of hydrogen, deuterium, helium and lithium; deuterium is chemically identical to hydrogen, but its nucleus consists of a neutron and a proton, whereas normal hydrogen only has one proton. The deuterium is destroyed when it becomes part of a star: at a temperature of around 500,000 K the proton and neutron pair in the nucleus are unstable. Lithium is another primordial element which cannot withstand the conditions inside stars. The normal processes of stellar evolution lead to hydrogen being steadily converted into helium. Explosive phenomena in the stars allow heavier elements to be synthesized from lighter ones in nuclear reactions. So, during its lifetime a galaxy will change in composition because some of the primordial elements are destroyed inside the stars, and they are replaced by heavier substances. As a galaxy evolves, therefore, we expect that it will become progressively richer in the total proportion of elements heavier than helium.

The major factors determining the morphological appearance of a galaxy of a given type are its age, the rate at which stars are forming and have formed in the past, and the masses of newly forming stars. Age is important because it determines how much cooking of the galactic material can have taken place inside stars. The rate of star formation also has an influence on the speed at which lighter elements are converted to heavier ones. The distribution of new stars by mass is of critical importance. This is because the matter which is used to make small stars, of a few solar masses or less, is essentially locked away for all time as far as galactic evolution is concerned: the lifetimes of the low mass objects are comparable to the age of the galaxies, and so they do not make a significant contribution to the enrichment of interstellar matter. On the other hand the O stars, which are massive, evolve in a few millions of years, which is short compared to a galaxy's lifetime, so the gas which goes into these stars is processed rapidly. Clearly then, the way in which the available matter for making stars is partitioned into stars of different mass will have an effect on galactic evolution.

By choosing an initial composition for galactic material, a star formation rate and a mathematical description of the mass distribution among new stars, it is possible to compute how a galaxy will evolve, and how its colour will change in time. Wallace Sargent and Leonard Searle have shown that the bluest spiral galaxies have had a constant rate of star formation for the last 10 billion years, whereas the reddest spirals have had a decreasing rate. It needs about 10^{10} years of evolution to mimic the real distribution of galactic colours if the mass function is the same as that in the solar neighbourhood.

One of the great tasks for the theory of galactic evolution is to explain why there exist both spiral and elliptical galaxies and to discover what relationship, if indeed there is any, links the two. The following speculative picture, which brings in the exploding galaxies too, can be sketched, although none of its details have been worked through thoroughly.

It is probable that the protogalaxies which condense from the primeval density fluctuations are basically elliptical in shape. Maybe all galaxies start as ellipticals. The argument then goes that the subsequent fate is governed by the interstellar gas in the protogalaxy. If giant explosions start at the galactic nucleus the gas could soon be swept out of the galaxy in a high-speed wind powered by shock waves propagating from the galactic core. Perhaps this type of behaviour can be identified with quasar and radio galaxy outbursts. Alternatively strong stellar winds, blown off hot stars, could cause expulsion of the gas. Galaxies that have lost the gas which did not immediately become locked into young stars remain as ellipticals for evermore, their stellar populations slowly ageing as time advances. In galaxies where explosions or winds do not blow out the interstellar material, the gas causes dynamical effects that lead to the galaxy collapsing from an elliptical into a disc. Once this has happened the gas remains trapped within the disc. These latter are the spiral galaxies.

What happens when we confront this broad-brush picture of evolution with the observations? Superficially the agreement is actually quite good. The most energetic radio sources are in fact identified with elliptical, and not spiral, galaxies, just as the model requires. Old elliptical galaxies are indeed relatively gas-free, except occasionally in the nuclei. Certainly they have none of the young stars that would form if significant amounts of gas were present. On the other hand the spirals are gas-rich, and we are postulating that it is just this gas which is responsible for the transition from elliptical to spiral galaxies. The picture has the one great merit that both major galactic types emerge from a common origin, with variations in the physical conditions in the protogalaxy accounting for the subsequent parting of the ways. Since the postulated physical mechanisms are actually observable and observed the crude model may well be on the right track.

13.4 Spiral structure

The study of the origin and evolution of spiral structure in galaxies centres on the following aspects: how does the structure arise, what keeps it going, and why is it not destroyed by the rotation of galaxies? Spiral galaxies are not rotating like rigid wheels: they display differential rotation, which means that the inner regions make a circuit of the galaxy in much less time than the outer regions. In our Galaxy, for example, gas clouds close to the galactic centres take only a few million years to orbit, whereas 250 million years must elapse before the material in the solar neighbourhood has made one rotation. The situation is similar in the solar system, where Mercury has an orbital period of only 88 days, whereas Pluto requires nearly 248 years. The differential rotation of material in spiral galaxies would tend to smear out the pattern after only one or two revolutions of the peripheries. However, one such revolution takes only 1 per cent or less than the age of the galaxy, so galaxies must use some physical trick to maintain the structure in the face of differential rotation.

One plausible suggestion to this problem was first suggested by B. Lindblad, and later developed in great detail at the Massachusetts Institute of Technology by C. C. Lin and his associates. Their basic idea is that the stars and gas in a galaxy move on roughly circular orbits in the galactic gravitational field, but the spiral pattern is due to a wave motion of such a nature that the pattern is semi-permanent and rotates as if it were a solid body. This wave motion concentrates stars and gas along two arms. As the arm advances it sweeps up new matter and discards stars from its trailing edge.

Connected with the density wave theory is the question of the driving mechanism for spiral structure: what keeps the waves going round and round the galaxy? Perhaps the barred spiral galaxies offer a clue. These central bars are observed to rotate as rigid bodies. It is possible, therefore, that this bar exerts a couple on the material further out and forces the density wave round the galaxy. It is most important to have some means of driving the spiral pattern, otherwise the waves die away in only a few revolutions.

An entirely different possibility is that spiral arms are created during close encounters between galaxies, and it is this idea which the Toomres have investigated. The general explanation is that tidal forces between galaxies will draw the material into arms. While this may account for the arm between M51 and NGC 5195, it is not an acceptable explanation for the majority of spirals. We know that structure will dissolve after only a few revolutions, so that all spirals would have had to have had encounters in the recent past. Too few of them are associated with sufficiently close, massive, companions for this idea to be correct.

Yet another approach to spiral structure has been made by theoretical astronomers who have large computers available. Their basic philosophy is to determine the evolution of an initially symmetrical disc of stars. To do this the individual star orbits within the gravitational field set up by all the other stars must be worked out. For a whole galaxy this would be a truly immense task. In order to reduce the computations systems of up to 100,000 stars have been considered. During the dynamical evolution of these star discs striking spiral patterns do arise. However, it appears that the conditions for this are rather artificial, and the theory has yet to attain wide recognition.

Ambartsumian has put forward a more far-reaching hypothesis than any of the above. He suggests that spiral arms arise from or are influenced by strong explosions in the nuclei of galaxies, particularly the explosions that result in the large-scale ejection of gas. This hypothesis proposes that spiral arm material is fired out of the rotating galactic nucleus.

Considerable observational evidence can now be brought to bear on the problem. In 1971 D. S. Mathewson, Pieter van der Kruit and W. N. Brouw published a high-resolution survey of the radio emission from M51. The most striking feature of their map is the clear delineation of two radio arms. The radio emission comes from gas, and they lie along the inner edges (or trailing edges) of the optical spiral arms, where they coincide with dust lanes. The coincidence of these radio arms with the dust lanes is claimed to be first-rate evidence that the spiral arms compress the interstellar gas, as expected on the density-wave theory of Lin. Radio emission is higher in the region of gas compression because the intensity of the magnetic field is greater.

Evidence in favour of a nuclear ejection model has come partly from observations within the Milky Way. These appear to show that the arms and ring structures in the nuclear region of our own Galaxy are expanding away from the centre, as if explosions had occurred there ten or so million years ago. More positive evidence has come from the mapping of the spiral galaxy NGC 4258 by van der Kruit, Jan Oort and Mathewson. In this galaxy the radio structure consists of two curved ridges which differ in shape and position from the optical arms. The inner parts of the radio arms also show up faintly on photographs taken in the red light of hydrogen, so they must contain at least some hydrogen gas. One interpretation of NGC 4258 is that the radio arms are ejected from the nucleus. An explosion may have taken place in this galaxy about 18 million years ago; the total ejected mass is about 10–100 million solar masses. Within a further 80 million years or so, the usual differential rotation effects will have caused the present arms to have evolved into a more conventional spiral structure. The mechanism of nuclear ejection could, therefore, be the means by which spiral structure is born and regenerated.

13.5 The scope of future research

As we have seen, galaxies have taught us a great deal about the universe at large. But there is still much to learn; in a few areas our knowledge is scanty. For example, the question of the existence of the intergalactic medium is still essentially unresolved. Perhaps the next generation of X-ray telescopes and instruments aboard the High Energy Astrophysics Observatory will gather data that bear on this and many other vital questions. The future evolution of the universe will be governed by the total mass that it contains; until we understand more about the regions between the galaxies we cannot know what that mass is. The relationship between the different morphological types of galaxy is not clearly delineated. Already we have asked why some galaxies are elliptical and others spiral. But where do the active galaxies—Seyferts, N–galaxies and compact galaxies—fit into the picture? What do quasars turn into as they age? Any complete description of the evolution of galaxies must address itself to these problems.

During the coming years we can expect a big push on the observational front. The 4-metre Anglo-Australian telescope in Australia became fully operational in early 1976. It now gives astronomers an instrument, rivalling the giant telescopes based in the northern hemisphere, with which to probe southern skies. Many new and interesting extragalactic results should emerge as the telescope picks its way through the Magellanic Clouds and the southern galaxies. More routine surveying of the southern heavens will continue to 1980 and beyond, with use being made of the British 1.2-metre Schmidt in Australia and the European Southern Observatory's Schmidt in Chile. By the end of 1976 most of the preliminary survey by the E.S.O. team will be complete. At this stage astronomers will have, for the first time, a photographic atlas of the whole sky complete to about the 20th magnitude. We may expect an increase in the numbers of strange galaxies, identified radio sources and quasars, for example, as work gets under way on the survey plates.

Further improvements in instrumentation will shortly lead to a dramatic increase in the amount of spectrographic data available for extragalactic objects. The use of electronic detectors, rather than photographic plates, and the increased application of digital processing techniques and on-line computer control are expected to revolutionize everyday observational astronomy. Spectra will be displayed directly onto a television monitor-screen rather than being recorded on a photographic plate. This type of approach should lead to higher efficiency and much greater productivity. Although the techniques can be applied to any celestial body, they are especially valuable for extragalactic studies, where very low light levels are encountered, and single observations currently take a long time.

The laboratory processing of photographic data should be greatly speeded up. By 1980 it is expected that entire Schmidt plates will be scanned in only a few hours by computer-controlled measuring devices. Instruments presently under construction will measure and analyse images automatically, providing accurate position data and morphological information. By processing groups of plates of the same area of the sky taken through different filters or at widely different times it will be possible to analyse the colours of images or to search for variable objects. This type of device will be extremely valuable in extragalactic studies because it will, for example, aid the compilation of complete catalogues of the various types of galaxies, and enable rapid analysis of galaxy clusters.

Increasing amounts of astronomical hardware will be launched into space in the coming years. The early X-ray astronomy detectors produced many surprises for students of extragalactic astronomy. Telescopes with better pointing accuracy and the ability to detect fainter sources will fill in our picture of the X-ray sky. Other high-energy radiations will be scrutinized. Perhaps it will not be long before large optical telescopes are operating in Earth orbit. They should, ideally, be controlled from the ground and telemeter all their data back. Above the turbulent atmosphere of Earth, they will provide much sharper images of galaxies than any ground-based telescope.

It is impossible to predict what problems will be solved by theoretical astronomers. It appears likely that they will continue to be attracted by such problems as: the origin and maintenance of spiral structures; the evolution of spiral and elliptical galaxies; the energy problem; the formation of galaxies and the nature of the early universe; high-energy phenomena in galaxies, galactic nuclei and black-hole astrophysics; and the nature of quasi-stellar objects.

Extragalactic astronomers face many challenges and problems, both observational and theoretical. An exciting future is in store, especially with the rush of southern hemisphere data.

Bibliography

The following is a short list of recent books that develop many of the topics which are discussed in this book.

B. J. BOK *and* P. F. BOK *The Milky Way*, Harvard University Press, Fourth Edition, 1974.

GEORGE B. FIELD *et al. The Redshift Controversy,* W. A. Benjamin, 1973.

J. S. HEY *The Radio Universe,* Pergamon Press, 1971.

FRED HOYLE *Astronomy and Cosmology*, W. H. Freeman, 1975.

DONALD H. MENZEL *et al. Survey of the Universe*, Prentice-Hall, 1970.

C. MISNER *et al. Gravitation*, W. H. Freeman, 1974.

P. J. E. PEEBLES *Physical Cosmology*, Princeton University Press, 1971.

HARLOW SHAPLEY *Galaxies*, Harvard University Press, Third Edition, 1972.

CHARLES A. WHITNEY *The Discovery of our Galaxy,* Angus and Robertson (UK) 1972; A. Knopf (USA) 1971.

Index of names and subjects

Abell, G., 54, 126
Allen, R., 101
Anderson, M. K., 121
aperture synthesis, 39
Arp, H., 49, 98, 100, 107, 163
association, stellar, 59, 61

Baade, W., 49, 51
Bahcall, J., 154
Barnard, E. E., 129
Becklin, E., 112
Bergh, S. van den, 54, 106, 128
Big Bang universe, 170, 188–91
black hole, 74, 162, 169, 176–7, 182–6
Bruno, G., 18
Burbidge, G., 71, 101, 124, 158–9
Burbidge, M., 71, 101, 158

Cepheid variables, 48–51, 65, 95, 96
colour index, 74
colour-colour diagram, 76
colour-magnitude diagram, 60–1, 64, 65
colours of galaxies, 74–6
containment problem, 107, 149–50
Copernicus, N., 18
cosmic rays, 176
cosmology, 125, 167–70, 187–98

Davis, R., 176
Democritus, 18
Dent, W. A., 157
distance modulus, 47, 60–1, 69
distance scale, 27–33, 44–58
Doppler effect, 53
Draper, H., 25
Dreyer, J., 23

Einasto, J., 169
electromagnetic spectrum, 34–6
electromagnetic waves, 36

Field, G., 173
flux unit, 137
Ford, K., 122
Fraunhofer, J. von, 25

galactic centre, 83, 110, 111–16
galactic cluster, *see* open cluster
galaxies, clusters of, 99, 125–32, 144, 173–4
galaxies, compact, 33
galaxies, dwarf, 126–8
galaxies, elliptical, 30, 31, 66–8, 73, 74, 90
galaxies, evolution of, 32, 108, 126, 187–98
galaxies, exploding, 105–7
galaxies, interacting, 97–108, 132
galaxies, irregular, 31–3, 66, 67, 73
galaxies, lenticular, 68
galaxies, N-type, 136
galaxies, radio, 133–50, 172, 178–86
galaxies, spiral, 30, 32, 66–8, 73, 74, 90–4
Galaxy, disc model, 21–9
Galileo, G., 18
Gaposchkin, S., 51
Ginzburg, V.L., 92
globular cluster, 59–61, 95, 128
gravitation, 161–2, 167–8, 177, 183–6
gravitational radiation, 115
Gunn, J., 164, 172

H I region, 81
H II region, 51, 84, 86, 88–90, 93, 100, 113, 118
Halley, E., 20
Hargrave, P., 106
Hartwick, F., 130

Hazard, C., 152–3, 164
helium, 81
Herschel, C., 23
Herschel, W., 21–6
Hertzsprung, E., 28
Hertzsprung-Russell diagram, 62, 64
Hey, J. S., 133
Hipparchus, 47
Hodge, P., 128
Holmberg, E., 73, 125
Hoyle, F., 164
Hubble, E., 29–33, 51, 125, 126, 129, 177
Hubble classification, 29, 30
Hubble constant, 167–9
Hubble diagram, 53–7
Hubble law, 54, 158, 160, 162–4
Huggins, W., 77
hydrogen, 21-cm line, 81

interferometer, 38–40, 134, 149, 151, 155
intergalactic medium, 52, 132, 150, 167–77
interstellar medium, 81–90, 110, 132, 155
interstellar molecules, 83, 88, 113–15
interstellar reddening, 76, 85

jansky, 137

Kant, I., 21, 24
Kapteyn, J., 27
Kelvin, Lord, 27
Kepler's Laws, 20
Kraft, R., 49, 50, 121
Kruit, P. van der, 112, 195–6

Laan, H. van der, 147
Leavitt, H., 28, 48, 49
light year, 45
Lin, C. C., 193
Lindblad, B., 193
Low, F., 120
luminosity function, 70
lunar occultation technique, 152
Lynden-Bell, D., 185
Lynds, R., 106, 107, 120

Magellanic stream, 104, 105
magnetic field in the Milky Way, 88
magnitude, 47
main sequence, 64
masses of galaxies, 70–4, 99, 169–70
Materne, J., 102
Mathewson, D. S., 195
Mattheus, T. A., 151
Méchain, P., 20
Messier, C., 20
Messier catalogue, 20
microwave background, 175, 188–90
Milky Way, 19–33, 93–6, 110–16, 126–9
Minkowski, R., 134
missing mass problem, 74, 168–77
Morgan, W., 32
moving cluster method, 46

nebula, 17–33
nebula, planetary, 26
neolithic astronomy, 18
Neugebauer, G., 112
neutrino, 176
neutron star, 74
Newton, I., 36
nova, 65, 96
nuclei of galaxies, 97, 105, 106, 107, 109–23, 132, 136, 149, 185

Oke, J., 121
Oort, J., 110–11
open cluster, 59–61
O stars, 88

Page, T., 73
Palomar Observatory Sky Survey, 42, 97, 125, 135
parallax, trigonometrical, 44–6
parsec, 45
Parsons, W. (Earl Rosse), 24
Penzias, A., 187
Peterson, B. A., 172
polarization, light, 87–8, 107
polarization, radio wave, 139–40, 174
Pooley, G., 91
populations, stellar, 79–80, 126
Prendergast, K., 71
pulsar, 181–2

quasar, 33, 151–66, 172–4, 178–86

radio contour map, 140–1, 161
radio source energy, 142, 178–86
Reber, G., 133
redshift, 153–4, 158–66, 171–2
Rees, M., 185

Rieke, G., 120
rotation curve, 71–3
RR Lyrae variables, 50, 65
Rubin, V., 122
Ryle, M., 133, 146

Salpeter, E., 90
Sandage, A., 32, 52–5, 106, 151
Sargent, W., 99, 100, 119, 121, 130, 193
Schmidt, M., 152–3
Sciama, D., 17
Scoville, N., 115
Searle, L., 193
Seyfert, C., 100, 118
Seyfert galaxies, 77, 119, 121, 122, 123, 135, 146, 148, 157
Shapley, H., 29, 49, 67, 127
Smith, F. G., 134
spectral index, 137–8
spectroscopy, 25, 41–2, 171
spectrum, absorption line, 26, 154–5
spectrum, emission line, 26, 77, 116, 119, 121, 134–6
spectrum, nebular, 26
spectrum, radio source, 41, 137–40
spectrum, stellar, 61–2
spinar, 181–2
spiral structure, 25, 82, 130
star formation, 88–91, 192
Stephan, M. E., 101
Stockton, A., 165
Stonehenge, 18
superclusters, 125, 126
supernova, 65, 66, 180, 181
synchrotron radiation, 107, 138, 178–180

Tammann, G., 53, 102
telescope, optical, 36–8
telescope, radio, 38–41
telescope, X-ray, 41
Terrell, J., 160
Third Cambridge (3C) Catalogue, 43
Toomre, A., 103
Toomre, J., 103
T Tauri stars, 88

UBV system, 75
Uhuru Catalogue, 43
Uranus, discovery of, 23

variability, 122–3, 154, 157–8
Vaucouleurs, G. de, 32
virial theorem, 74, 98–9
Vorontsov-Velyaminov, B. A., 97

Weber, J., 115
Walker, M., 118
white dwarf, 65, 74
white hole, 186
Wilson, R., 187
Windram, M., 146
Wright, T., 20

X-ray observations of galaxies, 130

Zel'dovich, Ya. B., 163
Zwicky, F., 33, 42, 98, 126, 173

Index of astronomical objects

Andromeda Nebula, *see* M31
Andromeda I-IV, 128
Arp 330, 98, 99

Centaurus A, 84, 143
Coma cluster of galaxies, 74, 125, 131–2, 148, 173
Crab Nebula, 120, 134
Cygnus A, 84, 133–42, 151, 179, 186

DA 240, 146, 178

Eta Carina nebula, 83

h and χ Persei, 94
Hyades, 59

IC 10, 126
IC 342, 126
IC 1613, 129
IC 1746, 163

Large Magellanic Cloud, 49, 50, 74, 94, 104, 126–8
Leo dwarf galaxies, 128
Local Group of galaxies, 125

Markarian chain of galaxies, 98
M31, 20, 26, 29, 50, 51, 66, 72, 79, 91, 93, 94, 105, 109, 116, 117, 126–30
M32, 117, 130
M33, 30, 51, 72, 73, 77, 91, 93, 94, 105, 109, 118, 126–8, 130
M51, 24, 25, 92, 103, 109, 117, 118, 195
M81, 117
M82, 105, 106, 135
M87, 54, 69, 105, 107, 135, 143
Maffei 1 and 2, 126

NGC 205, 130
NGC 566, 121
NGC 891, 92
NGC 1068, 119, 122
NGC 1097, 118
NGC 1221, 94
NGC 1275, 119, 120, 146–8
NGC 1365, 118
NGC 1886, 94
NGC 3227, 122
NGC 3516, 122
NGC 4038/39, 93
NGC 4051, 121, 122
NGC 4151, 119, 121, 122
NGC 4258, 92
NGC 4631, 92
NGC 4869, 131
NGC 4874, 131
NGC 5195, 103
NGC 5253, 66
NGC 6522, 79
NGC 6528, 79
NGC 6822, 129
NGC 7320, 101
NGC 7331, 101, 102
NGC 7603, 163

Orion nebula, 83, 113–14

Perseus cluster of galaxies, 98, 121, 147–8
PHL 957, 154–5
PHL 1226, 163
PKS 2251+11, 164
Pleiades, 59, 60, 94
Proxima Centauri, 45

Sagittarius A, 113–14
Sagittarius B2, 113–14
S Andromedae, 91
Seyfert's sextet, 100

Sirius, 45
Small Magellanic Cloud, 28, 48, 74, 94, 104, 126–8
Stephan's Quintet, 98, 101, 125

Virgo A (M87), 143
Virgo cluster of galaxies, 54, 74, 126, 130–2
VV 159, 99
VV 172, 98, 100
VV 282, 98

61 Cygni, 44–5

3C 9, 153, 155, 172
3C 33, 143, 150
3C 47, 155–7
3C 48, 151–3
3C 84, 146–7
3C 109, 123, 136
3C 120, 122
3C 123, 135
3C 129, 148
3C 147, 151–3
3C 196, 151–2
3C 231 (M82), 135
3C 236, 145–6, 178
3C 273, 151–3, 157
3C 274 (M87), 135
3C 279, 157
3C 295, 144
3C 371, 123, 136
3C 390.3, 123, 136
3C 465, 144

4C 05.34, 154–5
4C 11.45, 165
4C 11.50, 165
4C 24.23, 165
4C 26.48, 165

5C 4.81, 131, 148